Pipelined Analog to Digital Converter and Fault Diagnosis

Pipelined Analog to Digital Converter and Fault Diagnosis

Edited by
Alok Barua
Indian Institute of Technology, Kharagpur, India

IOP Publishing, Bristol, UK

ISBN 978-0-7503-1732-0 (ebook)
ISBN 978-0-7503-1730-6 (print)
ISBN 978-0-7503-1768-9 (myPrint)
ISBN 978-0-7503-1731-3 (mobi)

DOI 10.1088/978-0-7503-1732-0

Version: 20200301

IOP ebooks

British Library Cataloguing-in-Publication Data: A catalogue record for this book is available from the British Library.

Published by IOP Publishing, wholly owned by The Institute of Physics, London

IOP Publishing, Temple Circus, Temple Way, Bristol, BS1 6HG, UK

US Office: IOP Publishing, Inc., 190 North Independence Mall West, Suite 601, Philadelphia, PA 19106, USA

Dedicated to my elder brother Anjan Barua.

Contents

Preface

Mixed signal circuits incorporating both analog and digital systems have become increasingly prevalent in chip design. An analog system takes up a small area in a mixed signal circuit, however, its failure rate is rather high compared to the digital system. A systematic approach must be devised to fabricate a fault-free chip. The analog-to-digital converter (ADC) is a link between an analog signal and a digital circuit or system. The fastest ADC is the flash ADC, but it is hardware intensive. For example, a 4 bit flash ADC needs fifteen comparators as well as a priority encoder. The successive approximation ADC is the most popular for signal conditioning since it needs less hardware, although its conversion time is not very impressive. The pipelined ADC is a good compromise between speed and area-on-silicon. It has an in-built flash ADC of smaller dimensions. In today's world, the cost of testing the chip before it goes to the end user is extremely high. The only solution to this problem is to build a testing circuit on the chip itself so that testing can be performed without using any test equipment. This book has been written to describe the development of a custom-built pipelined ADC and the different techniques for its fault diagnosis using a built-in self-test system.

The book contains five chapters, with each chapter covering a pipelined ADC and a fault diagnosis technique. The diagnosis method is different in each of the chapters.

In chapter 1 we discuss a 1.8 V, 10 bit, 500 mega samples per second parallel pipelined ADC. Here the design of a high speed, low power, low voltage ADC in CMOS technology is described.

In chapter 2 we introduce the built-in self-test system where both the circuit and its diagnosis tool are implemented on the same chip. A built-in self-test system for a 1.8 V, 8 bit, 125 mega samples per second pipelined ADC is presented in this chapter.

The oscillation-based built-in self-test for ADC is a novel concept. The entire circuit is put into oscillation with some changes in its feedback system and it will produce an oscillation frequency that will guide us to detect whether the ADC is healthy or not. The design of an oscillation-based built-in self-test system for a 1.8 V, 8 bit, 125 mega samples per second pipelined ADC is covered in chapter 3.

The dynamic parameters are very important for any ADC design. In chapter 4 we focus on the evaluation of the dynamic parameters of a pipelined ADC with an oscillation-based built-in self-test system.

Recently, researchers have been interested in reconfigurable circuits, since this type of network is available in the soft core for a system-on-chip. The circuit can easily be reconfigured as and when it is needed by the chip designer, saving time and maximizing the yield of the chip. Chapter 5 is devoted to discussing reconfigurable built-in self-test architecture for a pipelined ADC. This chapter describes a new built-in self-test technique suitable for both functional and structural testing of analog and mixed signal circuits based on the oscillation-test methodology.

Alok Barua

December 2019, Kharagpur

Acknowledgments

I wish to thank Amal Saha who helped me to scan many hand drawn illustrations and prepare them for sending to the publisher.

I would like to acknowledge the support of my wife, Mausumi, for tolerating my absences while I spent many hours on this book over the last two years. I would like to thank my daughter, Arpita (Bibi), who is my continuous source of inspiration even though she has been staying in USA for more than a decade. Finally, I am grateful to my nonagenarian father, Mrinal Kanti Barua, who is the happiest person whenever I present him with any newly published book authored by myself.

Editor biography

Alok Barua

Alok Barua received his Bachelor of Technology in Instrumentation and Electronics Engineering, Master of Electronics and Telecommunication Engineering and PhD in Electrical Engineering from Jadavpur University and Indian Institute of Technology (IIT) Kharagpur in 1977, 1980 and 1992, respectively. In 1985 he joined the Department of Electrical Engineering, IIT Kharagpur and retired as full professor in June 2018. At present he is working as adjunct Professor in Indian Institute of Technology, Jammu. With more than thirty three years of teaching experience in IIT he has published many papers in his teaching and research areas—instrumentation, image processing, testing and fault diagnosis of analog and mixed signal circuit. He also holds a patent for the design of 'See Saw Bioreactor'. He had delivered invited lectures in many different universities of USA, Europe, South East Asia, Far East and Mediterranean countries. He worked as Visiting Professor/ Guest Professor/ Research Professor in University of Arkansas, USA, University of Karlsruhe, Frankfurt University, Yonsei University, Korea University and other institutions of the world. He co-authored the book *Computer Aided Analysis, Synthesis and Expertise of Active Filters* published by Dhanpat Rai and Sons, Delhi in 1995. He also co-authored the book entitled *Fault Diagnosis of Analog Integrated Circuit*, published by Springer, USA in the year 2005. He authored the book entitled *Fundamentals of Industrial Instrumentation*, and *Analog Signal Processing: Analysis and Synthesis*, published by Wiley India in 2011 and 2014, respectively. He also co-authored the book *Bioreactors: Animal Cell Culture Control for Bioprocess Engineering* published by CRC Press in 2015. He also co-authored two research monographs entitled *3D Reconstruction with Feature Level Fusion* and *Studies on Certain Aspects of a Newly Developed See Saw Bioreactor for animal Cell culture*, published by Lambert Academic Publishing, Germany in 2010 and 2011, respectively.

Contributor list

E Mallikarjun
National Institute of Technology Goa, India

Alok Barua
Indian Institute of Technology, Kharagpur, India

Md Tausiff
Bharat Heavy Electrical Limited EDN, Bangalore, India

Aniruddha Biswas
Steel Authority of India Limited Burnpur, India

Dhanunjay Nalla
Semtech, Neuchatel Switzerland

Raghavendra Singh Raghava
Texas Instruments. Bengaluru, Karnataka, India

Chapter 1

A 1.8 V, 10 bit, 500 mega samples per second parallel pipelined analog-to-digital converter

E Mallikarjun and Alok Barua

This chapter concentrates on pipeline architecture analog-to-digital converters (ADCs), which have become the architecture of choice for high speed and moderate to high resolution devices. The proposed pipelined ADC is used in analog preprocessing to divide the input signal range into sub-intervals and for amplification of a residue signal for further processing in the subsequent stages. The realization of the preprocessing stages is implemented using switched-capacitor (SC) circuits. In designing the current pipelined ADC, two major blocks are implemented using the SC technique. These blocks are the front end sample-and-hold amplifier (SHA) and multiplying digital-to-analog converter (MDAC). These circuits are implemented by amplifying the device at its core. The CMOS process is selected as the design platform for the current ADC, and the loads are capacitive in nature. Consequently, an operational transconductance amplifier (OTA) is preferred over op-amps as the amplifying device. In this chapter, a folded cascode OTA, SHA, MDAC and dynamic comparator are designed and simulated in Cadence using 180 nm technology.

1.1 Introduction

1.1.1 Motivation and goal

With the explosive growth of wireless communication systems and portable devices, the power reduction of integrated circuits has become a major problem. In applications such as personal communication systems (PCSs), cellular phones, camcorders and portable storage devices, low power dissipation, and hence longer battery lifetime, are essential. An example of a low power application is a wireless communication system. With the rapid growth of the Internet and information-on-demand, handheld wireless terminals are becoming increasingly popular. With the

doi:10.1088/978-0-7503-1732-0ch1

limited energy in a reasonably sized battery, minimum power dissipation in the integrated circuit is necessary.

With the rapid growth of the information superhighway, large amounts of data are stored in storage devices and accessed frequently. In order to transmit a large amount of data in a short period of time, a high transfer rate in storage devices is required. This translates directly into a higher data conversion rate in the read channels of magnetic storage devices, such as in an SCSI hard drive. However, in order to achieve even higher transfer rates for some multimedia applications, the speed of the analog-to-digital converter (ADC) needs to be improved.

To achieve the goals mentioned above (i.e. low power, low voltage and high speed), CMOS technology is very attractive for several reasons. First, its low cost and high integration level have made CMOS technology superior to bipolar technology. Because of this, several low power CMOS design techniques have been developed. With scaled CMOS technology, the high speeds which were once reserved for bipolar or other fast transistors can be achieved.

With the above motivations, the goal of this research is to build a high speed, low power, low voltage ADC in 0.18 μm CMOS technology.

1.1.2 Chapter organization

This chapter is organized as follows. In section 1.2, pipelined ADC architectures are discussed. First, the evolution of the pipelined ADCs is presented. Then a detailed description of pipelined ADC architecture is provided with an emphasis on the advantages of pipelined ADC architecture for low power and high speed. In section 1.3, the main building blocks of pipelined ADCs and folded cascode operational transconductance amplifiers (OTAs) are designed. The requirements of gain and unity gain bandwidth are discussed along with gain boosted OTAs. The common mode (CM) feedback circuits are also presented here. In section 1.4, noise issues in the sample-and-hold (S&H) circuit are discussed, as are charge injection error, bottom-plate sampling, the requirements for bandwidth and simulation results. In section 1.5, the operation of a multiple digital-to-analog converter (MDAC) is presented along with the requirements for bandwidth limitations of MDACs. In section 1.6, an introduction to the dynamic comparator and its operation along with simulation results are presented. In section 1.7, we provide our conclusions and future work is suggested to complete the circuit design.

1.2 Pipelined analog-to-digital converter architecture

1.2.1 Evolution of pipelined ADC architecture

Since the emergence of digital signal processing, ADCs have played a very important role, interfacing between the analog and digital worlds. They digitize the analog signals at a fixed time period (frequency). The time period is generally specified by the application. This time is calibrated using the Nyquist sampling theorem which states that a band limited signal having no spectral components above f_m Hz can be determined uniquely by values sampled at uniform intervals of T_s seconds, where

$$T_s \leqslant \frac{1}{2f_m}.$$

This condition needs to hold in order to reconstruct the original analog signal completely. Since algorithms can be implemented very inexpensively in the digital domain and if the samples acquired satisfy the Nyquist sampling theorem, signals can be reconstructed perfectly after the digital signal processing. Hence, the ADC acts as a bridge between two domains and its accuracy is highly critical for the performance of the system.

1.2.2 Flash architecture

The most straightforward way to perform the analog-to-digital conversion [1] is to compare the sampled analog signal with different reference levels. Figure 1.1 shows a conceptual diagram for such a converter. The input signal is first sampled by the S&H circuit, and during the hold cycle the comparators make a decision as to whether or not the sampled value is greater or smaller than the reference voltages.

The output data are then collected and construct a digital representation for the sampled analog signal. Because of the direct comparison, each reference level needs to be one least significant bit (LSB) apart from each other. Assuming that an N bit ADC is desired, the number of comparators required is $2^N - 1$. Further assume that

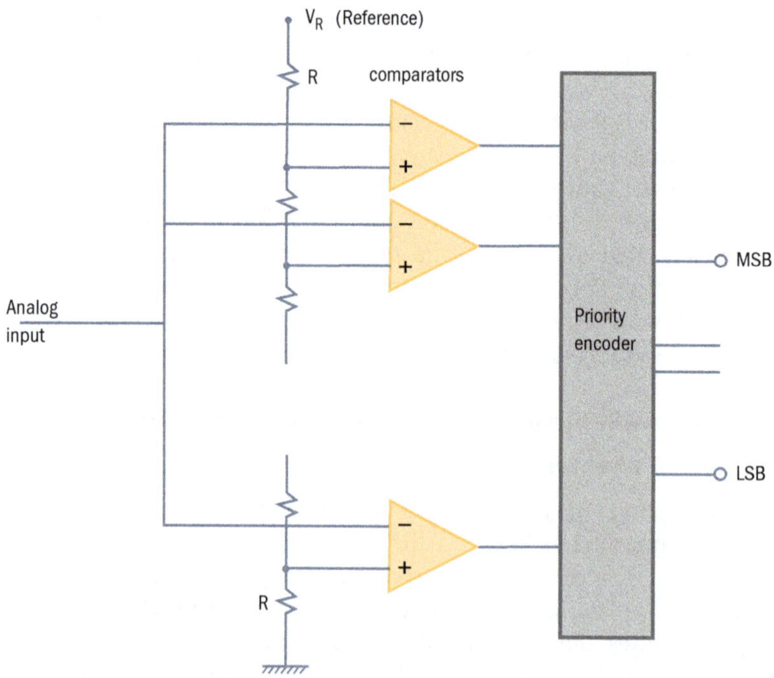

Figure 1.1. A conceptual diagram of a flash ADC.

the full-scale input is 1 V, and then the LSB size is 2^{-N}. Therefore, the offset of the comparator needs to be much less than this value, typically for 10 bit resolution it is less than 1 mV. In CMOS technology, this offset requirement is difficult to achieve. Some special circuit techniques are required to reduce the offset of the comparator. However, these techniques require more power and hence may not be practical. Therefore, this family of converters is limited to 8 bit or lower resolution. The name flash converter is given to this architecture because of its fast conversion rate, which takes a single clock cycle to perform the conversion.

1.2.3 Two-step flash architecture

The two-step flash architecture can avoid the problem of a large number of low offset comparators that is present in single-step flash architecture. Figure 1.2 shows the block diagram of a two-step flash ADC which reduces the comparators to $2^{\frac{N}{2}+1}$. Similar to flash converters, the analog input is first sampled by the S&H circuit, and during the hold period the first flash ADC performs a coarse quantization on the hold signal. Next the hold signal is subtracted from the output of the DAC, and finally the residue of the subtraction is passed down for fine quantization to obtain the full resolution of the converter. Although this architecture still requires a low offset comparator with the full resolution of the converter, the number of low offset comparators required is reduced significantly. Only $2^{\frac{N}{2}}$ coarse comparators are required in the first half of the converter, therefore the overall number of comparators used is reduced. By using concurrent processing, the throughput of this architecture can sustain the same rate as flash ADC. However, the converted outputs have a latency of two clock cycles due to the extra stage to reduce the number of precision comparators.

The advantage of this architecture is its low count on precision comparators and it will lead to lower power. The throughput is the same as that of flash converters because of the concurrent processing of signals; however, an extra clock cycle is required because it requires two steps to complete the conversion. If the system can

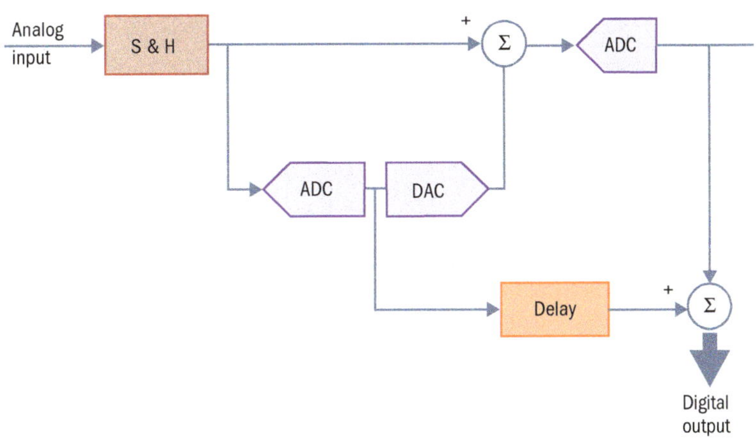

Figure 1.2. Block diagram of a two-step flash ADC.

tolerate latency of the converted signal, two-step flash is a low power, smaller area alternative. The disadvantage is that both the subtraction and precision comparators still need to be the full resolution of the ADC. As mentioned above, it is very difficult to achieve resolution above 8 bit in CMOS transistors without special techniques to compensate for the offset. Subtraction accuracy can be relaxed by using a wider range of precision comparators in the second stage, i.e. digital correction. Interstage gain can be used here to tolerate a larger comparator offset for the second-stage precision comparators.

1.2.4 Conventional pipelined ADC architecture

In the two-step flash converter, an additional amplifier can be used to relax the comparator offset in the second stage. In the same way, if we amplify the subtracted residue signal from the first stage to the full scale, the offset requirement of the second-stage comparators can be relaxed. Figure 1.3 shows a two-step flash converter with an additional amplifier with gain A. However, the gain of the amplifier needs to be carefully designed according to the first stage resolution and the overall ADC resolution. As an example, a 10 bit, two-step flash ADC utilizes a switched-capacitor (SC) amplifier. The amplifier is required to settle 10 bit resolution in approximately half the clock period, and with a gain of $2^5 = 32$. To meet this requirement in an SC circuit at high speed is very difficult and may take a lot of power, mainly due to the small feedback factor [2].

In order to reduce the power further, the per-stage resolution is to be reduced and a greater number of these stages are to be cascaded to achieve the full resolution. This architecture is called the pipelined architecture and is shown in figure 1.4. The sampled input signal is converted to the resolution of the stage, B bits; concurrently it is also subtracted from the DAC output of the present stage of digital output. The residue is then amplified by the factor 2^B and passed down to the next stage. Identical operation is performed for each stage and the digital outputs are combined properly to achieve the required $M \times B$ bit full ADC resolution. Here M is the number of stages and B is the number of bits in each stage.

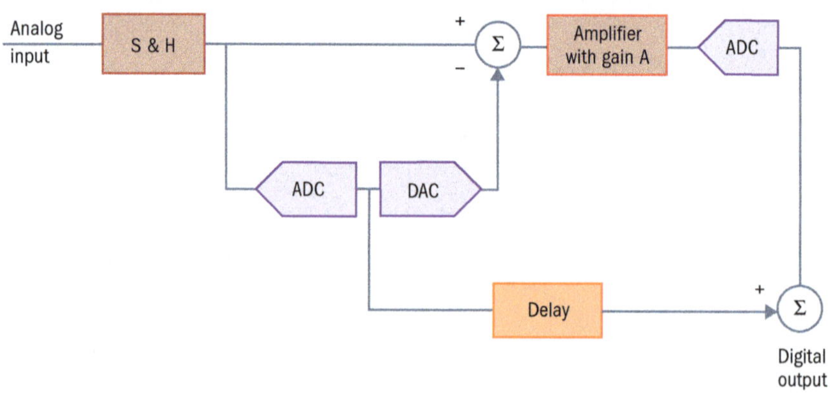

Figure 1.3. Block diagram of a two-step flash ADC with an additional amplifier with gain.

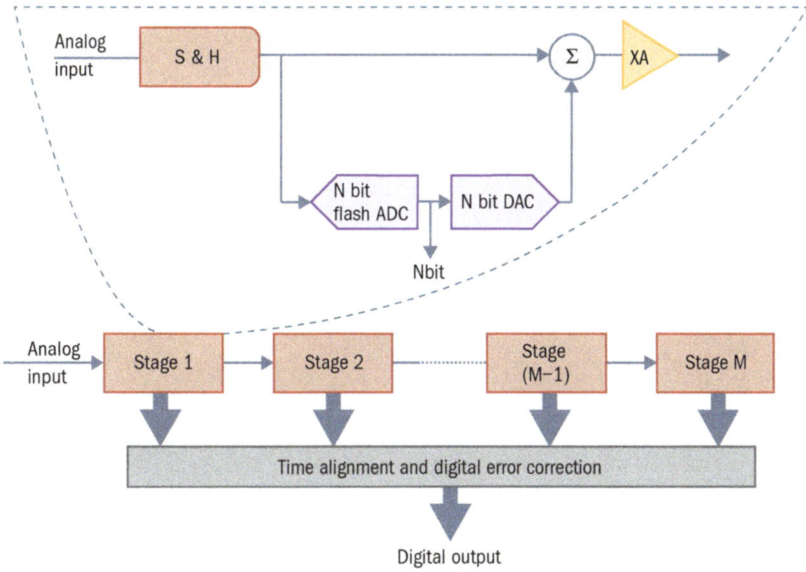

Figure 1.4. Block diagram of a conventional pipelined ADC.

The advantage of this architecture is its reduced complexity compared to the huge number of comparators needed in flash ADC. Moreover, it provides a good compromise between the speed and component count or circuit complexity. An ADC with full resolution can be achieved by cascading an appropriate number of identical pipelined stages with a given per-stage resolution. Therefore, the hardware cost is a linear function of resolution, if all the requirements are met. Some capacitor trimming techniques may be required to correct for the SC circuit gain and non-ideal subtraction (capacitor mismatch). With concurrent processing (interleaving between stages), the throughput achieved is the same as in the flash case. The major disadvantage of this architecture is the latency in the converter. Time alignment and digital error correction are needed in this ADC.

1.2.5 Redundancy and digital correction

To build pipelined ADCs with a large tolerance to component non-idealities, redundancy is introduced by making the sum of the individual stage resolutions greater than the total resolution [3, 4]. When the redundancy is eliminated by a digital-correction algorithm, it can be used to eliminate the effects of analog-to-digital sub-converter (ADSC) nonlinearity and interstage offset on the overall linearity. Figure 1.5 shows a block diagram of one stage in a pipelined ADC with offsets in series with both the ADSC and the DAC. A 2 bit stage is used as a representative example. The magnitudes of the offsets are both equal to 1/2 LSB at a 2 bit level ($E_r/4$, where $\pm E_r$ is the full-scale range of the ADC). Figure 1.6 shows a plot of the ideal residue versus held input without the offsets, and figure 1.7 shows the same plot with the offsets. The ADSC offset uniformly shifts the locations of the decision levels to the right, and the DAC offset shifts the entire plot down. Because

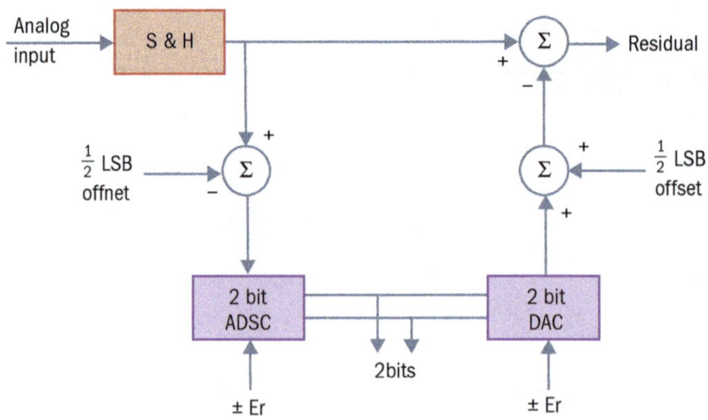

Figure 1.5. Block diagram of one stage with offsets in both the ADSC and the DAC.

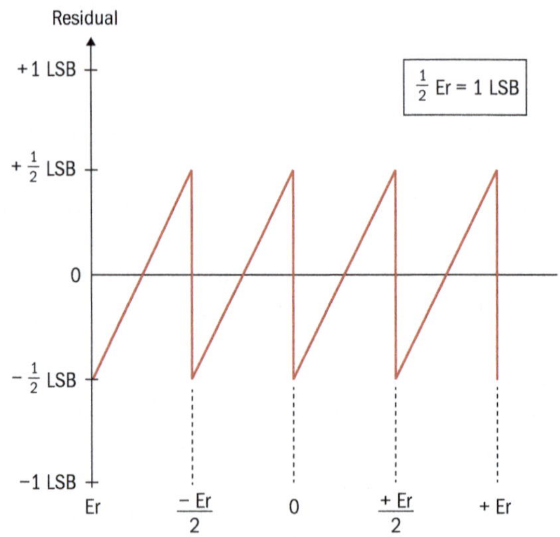

Figure 1.6. Ideal residual versus held input (without offsets).

there are no decision levels at half scale in figure 1.7, a multi-stage ADC using stages with this transfer characteristic will inherently have excellent linearity at half scale.

Let the correction range be defined as the amount of decision-level movement that can be tolerated without error. If the DAC and S&H amplifier (SHA) are ideal and the additional amplifier gain is 2, the amplified residue from figure 1.8 remains within the conversion range of the next stage when ADSC nonlinearity shifts the decision levels by no more than ±1/2 LSB at a 2 bit level. Under these conditions, errors caused by the ADSC nonlinearity can be corrected; therefore, the correction range here is $\frac{1}{2}$ LSB at a 2 bit level or $\pm E_r/4$, which means that the ADSC linearity must only be commensurate with the stage resolution instead of with the entire ADC resolution. Furthermore, because the offset introduced into the ADSC in figure 1.8 shifts the decision levels to the right by the amount of the offset, the digital output is

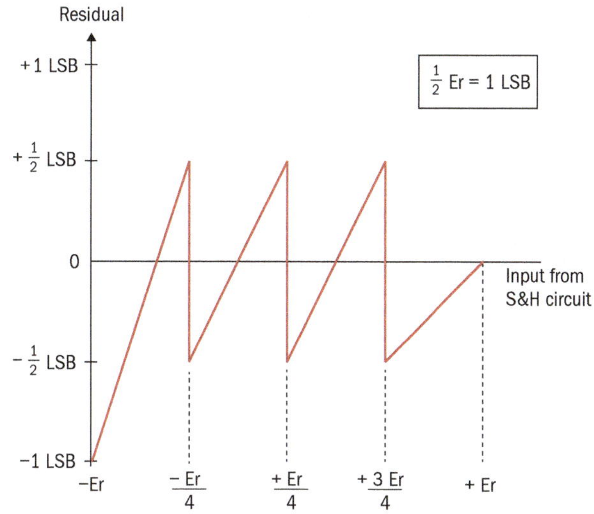

Figure 1.7. Ideal residual versus held input (with offsets).

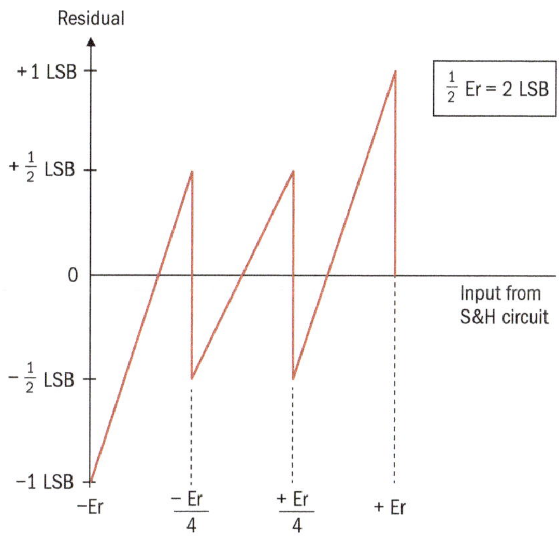

Figure 1.8. Ideal residual versus held input without top comparator.

always less than or equal to its ideal value if ADSC nonlinearity can shift the decision levels back to the left by no more than this amount. Thus, the correction here requires either no change or addition. Because the associated digital-correction logic does not perform any subtraction, it is easier to test than the conventional correction logic.

Removal of the top comparator in each stage forces a correctable error to occur in the uncorrected ADSC output for a full-scale input. Figure 1.9 shows the new ideal residue plot for the 2 bit case. Without the top comparator, the digital output never

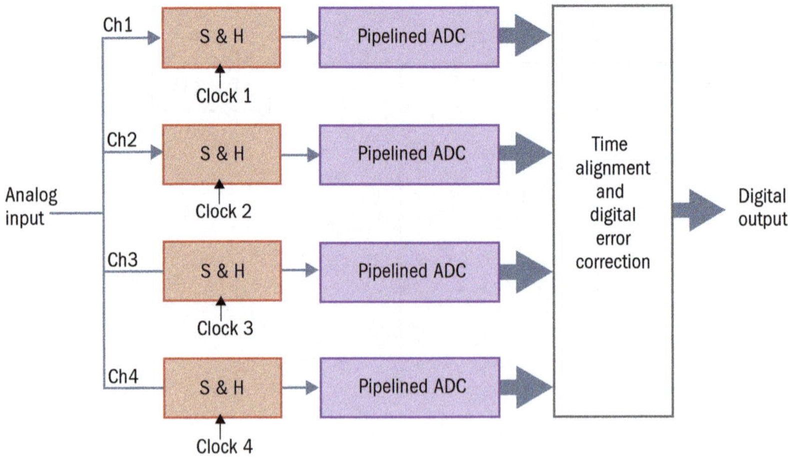

Figure 1.9. Parallel pipelined architecture.

reaches code 11 and the residue continues to rise for increasing inputs greater than $\frac{1}{4}$ of the reference. Because the resulting residue on the right side of figure 1.9 has the same magnitude as the left-end residue in figure 1.8, removal of the top comparator does not increase the magnitude of the maximum residue. To obtain code 11 from this stage after correction, the correction logic must increment the output of this stage. Furthermore, to obtain code 00 from this stage after correction, since the correction logic cannot subtract, it must do nothing. This simplifies the testing of the digital-correction logic. Furthermore, after removal of the top comparator, the correction range is still $\pm\frac{1}{2}$ LSB at a 2 bit level because the remaining decision levels can move by this amount before the resulting residue exceeds the conversion range of the next stage. Therefore, in this example, only two comparators are needed in each stage except the last. The last stage still needs three comparators because its output cannot be corrected. In general, if n is the number of digital output bits per stage, while the last stage needs $2^n - 1$ comparators, every other stage only needs $2^n - 2$ comparators. As a result, the resolution of each stage except the last is $\log_2(2^n - 1)$ bits. If $n = 2$ bits, as in this example, the resolution per stage is about $1\frac{1}{2}$ bits.

1.2.6 The parallel pipelined ADC

Because of the higher precision and higher sampling rate compared to other kinds of ADC, the pipeline ADC is used extensively in the realm of high speed digital transmission, digital image processing, etc. With an individual pipelined ADC it is difficult to accomplish such high speed performance. To achieve higher sampling rates, a parallel pipeline is suitable. Figure 1.9 shows a parallel pipelined architecture with a four channel which operates based on a time-interleaved technique [5]. To achieve 500 MSPS and 10 bit resolution, each channel performs 125 MSPS and 10 bit resolution. The operation of each channel is shown in figure 1.10 with respect

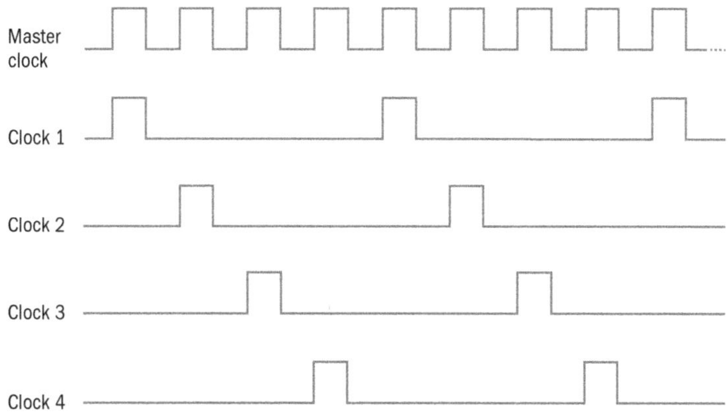

Figure 1.10. Clocks for the operation of each pipelined ADC channel with time.

to time (t). All the clocks in the S&H circuits are non-overlapping and they are derived from a single master clock. The main drawback of parallel pipelined ADC is the huge power consumption compared to single channel pipelined ADC. We can reduce the power consumption to some extent using an amplifier sharing technique [6].

1.3 Operational transconductance amplifier (OTA)

1.3.1 Introduction

An op-amp or an OTA is the main building block of any SC circuit. Therefore, the high frequency performance and dynamic range of this device is of great importance. Moreover, it consumes most of the dc power of the complete design. For the complete design of the pipelined ADC, the specifications of OTA, i.e. gain, output impedance and gain bandwidth product, will greatly influence the overall performance of the circuit.

1.3.2 Requirement of gain

Figure 1.11 shows a closed loop amplifier. It is also called a feedback amplifier. In closed loop operation, the dc gain of the amplifiers depends on the forward path gain (A) and feedback factor (β). For ideal operations, the forward path gain must tend to an infinite value. However, in the practical case the gain is finite. Thus the designed amplifier gain error must be less than the tolerable error in the ADC. The gain requirement for the 10 bit operation should be more than 60.2 dB, and is described using the following equations.

Closed loop gain:

$$A_{\text{closed}} = \frac{A}{1 + A\beta}.$$

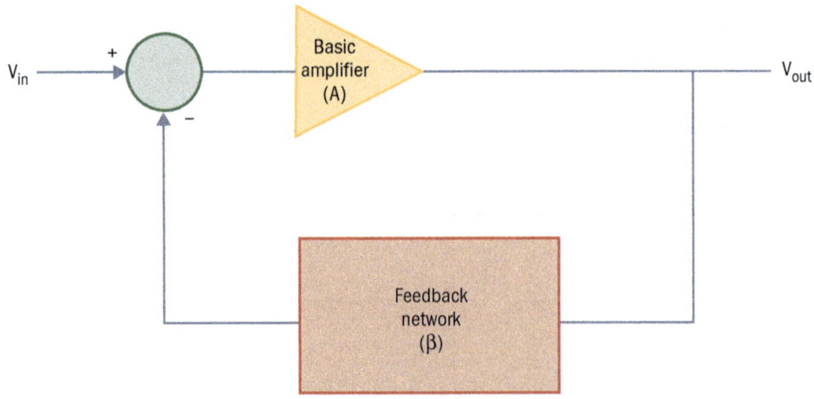

Figure 1.11. A feedback amplifier.

Consider $\beta = 1$, then

$$A_{\text{closed}} = \frac{A}{1 + A} \Longrightarrow \frac{1}{1 + \frac{1}{A}}$$
$$\approx 1 - \frac{1}{A}.$$

For proper ADC operation, the gain error $(\frac{1}{A})$ must be less than $\frac{1}{2}$ LSB of the ADC [7]:

$$\Rightarrow 1/A < 1/2^{N-1} \quad (N = \text{total number of bits})$$
$$\Rightarrow 20 \log(\text{A}) > 6.02 \cdot N \text{ dB}. \tag{1.1}$$

For 10 bit operation the required gain should be more than 60.2 dB.

1.3.3 Requirement of unity gain bandwidth (UGB)

The small signal model for an SC amplifier is shown in figure 1.12 [7], where the OTA is represented with a single-pole model consisting of g_m, g_o and C_l. The capacitance C_l is the sum of the OTA output capacitance and the external load capacitance (MDAC input capacitance, comparator input capacitance). The capacitance at the OTA input is represented by C_{in} and the output conductance by g_o. This circuit is used to find the settling time constant using a pulsed current source i_i as the excitation signal. The small signal analysis yields the following transfer function:

$$\frac{v_0}{i_i} = \frac{g_m - sC_f}{s[(g_m C_f + g_o C_f + g_o C_{i,\text{tot}}) + s(C_{i,\text{tot}} C_f + C_{i,\text{tot}} C_l + C_l C_f)]} \tag{1.2}$$
$$C_{i,\text{tot}} = C_s + C_{\text{in}}.$$

The output voltage for a current impulse with total integrated charge Q can be written (in partial fraction form) as

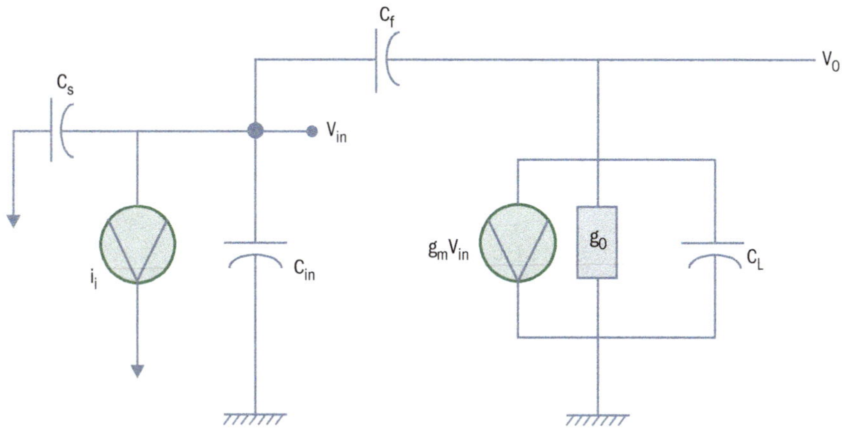

Figure 1.12. The small signal model for an SC amplifier in hold mode.

$$v_o = \frac{g_m Q}{s(g_m C_f + g_o C_f + g_o C_{i,\mathrm{tot}})} - \frac{\dfrac{C_F}{C_{i,\mathrm{tot}} C_f + C_{i,\mathrm{tot}} C_l + C_l C_f} + \dfrac{g_m}{g_m C_f + g_o C_f + g_o C_{i,\mathrm{tot}}}}{\dfrac{g_m C_f + g_o C_f + g_o C_{i,\mathrm{tot}}}{C_{i,\mathrm{tot}} C_f + C_{i,\mathrm{tot}} C_l + C_l C_f} + s} Q \tag{1.3}$$

$$v_o = \frac{Q}{s\left(C_f + \frac{g_o}{g_m} C_{i,\mathrm{tot}} + \frac{g_o}{g_m} C_f\right)} \left(1 - e^{-\frac{g_m C_f + g_o C_f + g_o C_{i,\mathrm{tot}}}{C_{i,\mathrm{tot}} C_f + C_{i,\mathrm{tot}} C_l + C_l C_f} t}\right) - \frac{C_f Q}{C_{i,\mathrm{tot}} C_f + C_{i,\mathrm{tot}} C_l + C_l C_f} \left(e^{-\frac{g_m C_f + g_o C_f + g_o C_{i,\mathrm{tot}}}{C_{i,\mathrm{tot}} C_f + C_{i,\mathrm{tot}} C_l + C_l C_f} t}\right) \tag{1.4}$$

from where the settling time constant can be identified as

$$\tau = \frac{C_{i,\mathrm{tot}} C_l + C_{i,\mathrm{tot}} C_l + C_l C_f}{g_m C_l + g_o f + g_o C_{i,\mathrm{tot}}}. \tag{1.5}$$

Since g_m is large compared to g_o, the time constant can be approximated with

$$\tau \approx \frac{C_{i,\mathrm{tot}} C_f + C_{i,\mathrm{tot}} C_l + C_l C_f}{g_m C_f}. \tag{1.6}$$

A well-known relation states that the gain bandwidth product of a single-pole OTA is

$$A_0 \alpha = \frac{g_m}{2\pi C_l},$$

where A_0 is the open loop gain and α is the low frequency pole:

$$\tau = \frac{1}{2\pi A_0 \alpha}\left(1 + \frac{C_{i,\text{tot}}(C_l + C_f)}{C_l C_f}\right). \tag{1.7}$$

This time constant must be below $\frac{\tau_S}{2}$ and the error value must be below $\frac{1}{2^N}$ (LSB):

$$\Rightarrow A_0\alpha \geqslant \frac{2\left(1 + \frac{c_{i,\text{tot}}(C_l + C_f)}{C_l C_f}\right) \cdot \ln(2) \cdot N}{2\pi\tau_s}. \tag{1.8}$$

For $N = 10$, $C_{i,\text{tot}} \approx C_f$, $C_l \approx 2 \cdot C_f$:

$$\Rightarrow A_0\alpha \geqslant 5.5\ \tau_s.$$

1.3.4 Folded cascode OTA

The folded cascode OTA, shown in figure 1.13, is probably the most commonly used architecture in SC circuits [8, 9]. It is a fully differential circuit with differential input and differential output. Moreover, it provides a larger output swing and input CM

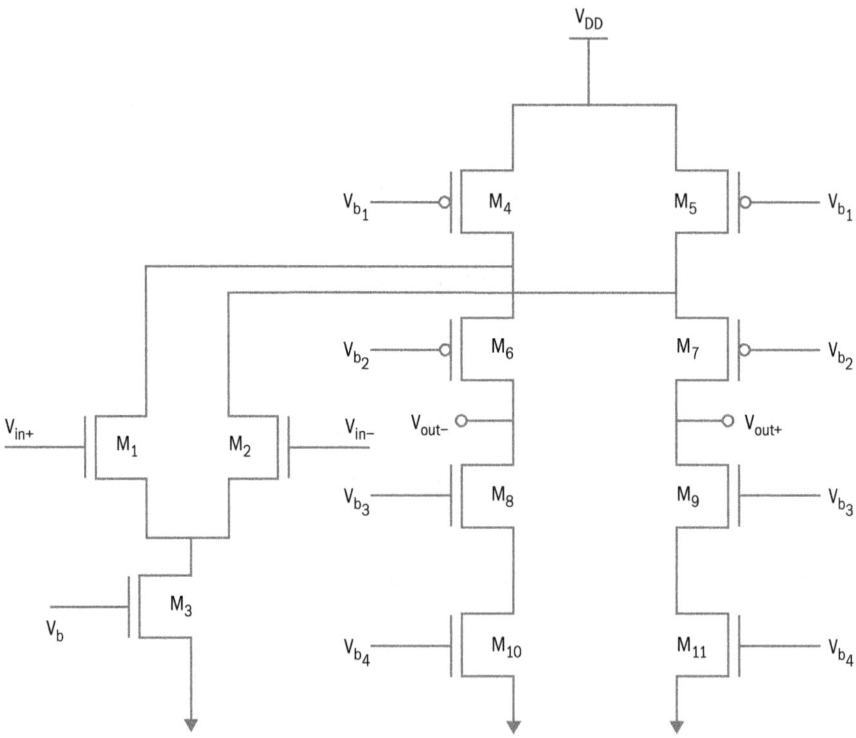

Figure 1.13. Folded cascode operational transconductance amplifier.

range compared to other single-stage OTAs (telescopic OTAs). The biasing voltage V_b and V_{b_1} to V_{b_4} are supplied from a beta multiplier self-biased reference circuit.

The choice between the NMOS and PMOS input pair has to be made on the basis of the required phase margin. The NMOS input architecture, shown in figure 1.13, offers a large gain bandwidth ($A_0\alpha$) product:

$$A_0\alpha = \frac{g_{m_1}}{c_l}, \tag{1.9}$$

g_{m_1} = transconductance of input NMOS transistors;
c_l = load capacitance at the output node.

In this folded cascode architecture, the upper limit of input CM range (ICMR) is governed by M_4 and M_5 PMOS transistors, and the lower range is limited by the current source M_3:

ICMR (upper) = $(V_{DD} - V_{D_{sat_4}})$;
$V_{D_{sat_4}}$ = drain saturation voltage of M_4 or M_5;
V_{Th} = threshold voltage of M_1 or M_2;
ICMR (lower) = $(V_{GS_1} + V_{D_{sat_3}})$;
$V_{D_{sat_3}}$ = drain saturation voltage of transistor M_3.

The upper limit of output CM range (OCMR) is limited by M_5 and M_7 (M_4 and M_6), the lower range is limited by M_9 and M_{11} (M_8 and M_{10}):

OCMR (upper) = $V_{DD} - V_{D_{sat_5}} - V_{D_{sat_7}}$;
OCMR (lower) = $V_{D_{sat_9}} + V_{D_{sat_{11}}}$.

The overall gain of the folded cascode depends on the output impedance (R_{out}):

$$R_{out} = (g_{m_7} \cdot r_{DS_7}(r_{DS_5} \parallel r_{DS_2})) \parallel (g_{m_9} r_{DS_9} \cdot (r_{DS11})).$$

1.3.5 Gain boosted folded cascode OTA

In many applications the op-amp dc gain requirement is higher than the achievable gain with the simple single-stage topologies. Techniques to enhance the op-amp dc gain without going into multi-stage architectures are desired in high speed circuits, where the high current levels make the transistor g_{DS} large. A very widely used method is based on improving the cascoding effect of a single MOS transistor by using local negative feedback [10]. The resultant circuit is often referred to as a regulated cascode and it is implemented in a current source that is shown in figure 1.14. In this, the auxiliary amplifier encloses the cascode transistor M_2 in a feedback loop, making the voltage on its source node almost constant. As a result, the output impedance of the current source is given by

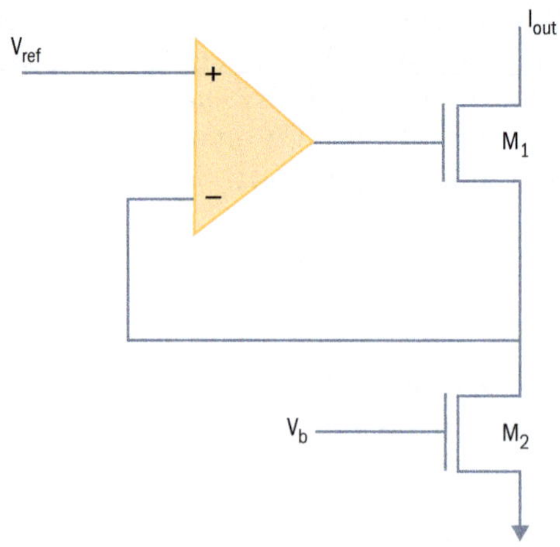

Figure 1.14. Regulated cascode current source.

$$r_{\text{out}} = \frac{A_1 g_{m_1} + g_{m_2}}{g_{\text{DS}_1} g_{\text{DS}_2}},$$

where g_{m_1} and g_{m_2} are the transconductances of transistors M_1 and M_2, respectively. g_{DS_1} and g_{DS_2} are the output conductances of transistors M_1 and M_2, respectively.

The regulated circuit has the improvement of the impedance by the gain of the regulation amplifier A_1 and when the current source is utilized in an OTA the DC gain is increased by the same amount. A fully differential (differential input and differential output) folded cascode OTA with a gain boosting stage is shown in figure 1.15. The total output impedance of the gain boosted stage is given by

$$R_{\text{out}} = (A \cdot g_{m_7} r_{\text{DS}_7} \cdot (r_{\text{DS}_1} \parallel r_{\text{DS}_2})) \parallel (A \cdot g_{m_9} \cdot r_{\text{DS}_9} \cdot (r_{\text{DS}_{11}})). \tag{1.10}$$

1.3.6 Frequency response of a folded cascode OTA

Any typical folded cascode structure has two poles 0. Since the folded cascode amplifier mostly has a capacitive load, its dominant pole is created by the load capacitance C_l and the output resistance R_{out} seen from the output node:

$$R_{\text{out}} = ((g_{m_7} \cdot r_{\text{DS}_7} \cdot (r_{\text{DS}_7} \parallel r_{\text{DS}_2}))) \parallel (g_{m_9} r_{\text{DS}_9} \cdot (r_{\text{DS}_{11}})) \tag{1.11}$$

$$p_1 = \frac{1}{[(g_{m_7} \cdot r_{\text{DS}_7} \cdot (r_{\text{DS}_7} \parallel r_{\text{DS}_2})) \parallel (g_{m_9} r_{\text{DS}_9}(r_{\text{DS}_{11}}))]C_l}. \tag{1.12}$$

The non-dominant pole is generated by the cascode mode of the OTA, i.e. at the drains of transistors M_4 and M_5. The parasitic capacitance C_p formed at this node and the resistance seen from the drain of transistor M_5 forms this pole,

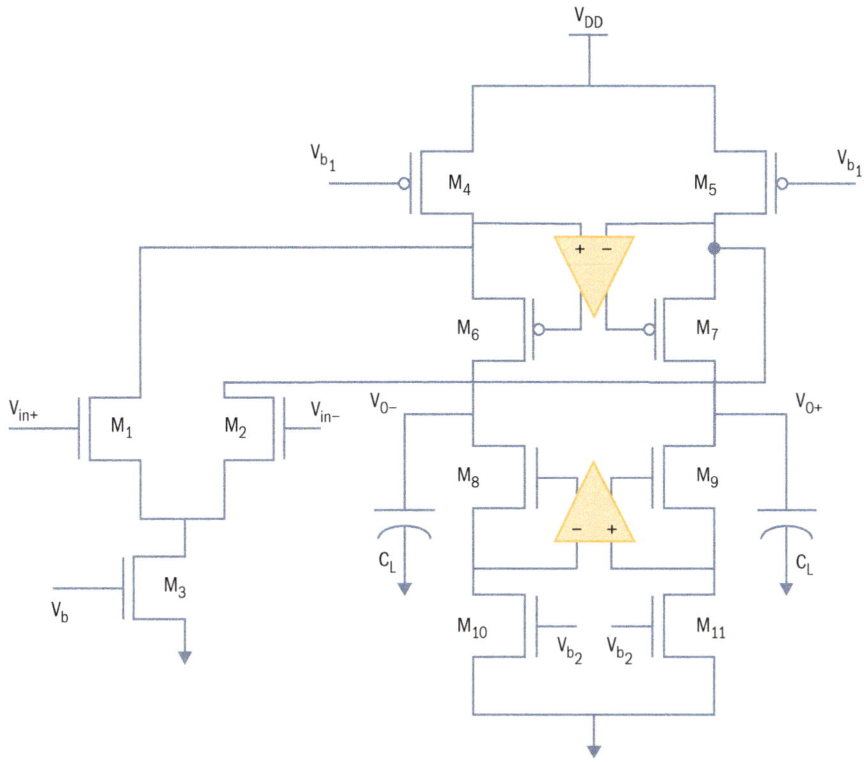

Figure 1.15. Fully differential folded cascode OTA with a gain boosting stage.

$$p_2 = \frac{1}{R_{11}C_p} \tag{1.13}$$

$$R_{11} = \left(\frac{g_{m_9}\, r_{\mathrm{DS}_9}\, r_{\mathrm{DS}_{11}}}{g_{m_7}\, r_{\mathrm{DS}_7}}\right) \parallel (r_{\mathrm{DS}_5} \parallel r_{\mathrm{DS}_2}). \tag{1.14}$$

The parasitic capacitances are formed by the gate-to-source and bulk-to-source capacitance of M_7 and gate-to-drain and drain-to-bulk capacitance of M_5. The folded cascode OTA should ideally show a single-pole behavior, since it is a single-stage OTA. Thus, the non-dominant pole P_2 must occur at a much higher frequency far from the unity gain bandwidth (f_u). For the fast settling time and stability, the phase margin of OTA $\geqslant 60°$. Therefore, the non-dominant pole is placed at $\sqrt{3}f_u$ [11].

1.3.7 Results

Simulation results of the gain boosted folded cascode OTA are determined at the typical temperature, voltage and process corners. Table 1.1 presents OTA specification and target values and achieved values. Figure 1.16 presents the magnitude and phase plot of the folded cascode OTA.

Table 1.1. OTA specifications; targeted achieved values.

Specifications	Target	Achieved
Gain	60 dB	69.78 dB
Unity gain bandwidth (f_u)	600 MHz	770 MHz
ICMR	300 mV	300 mV
OCMR	300 mV	300 mV
Slew rate	400 V μs^{-1}	572 V μ^{s-1}
Settling time	4 ns	0.418 ns
CMRR	100 dB	126.07 dB
Power consumption	As small as possible	3.26 mW

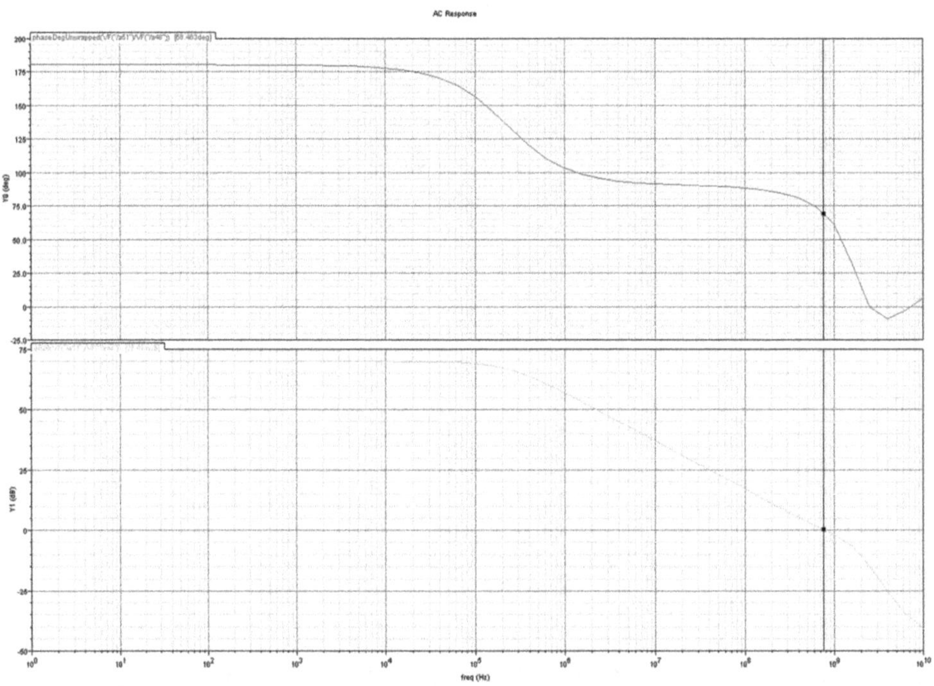

Figure 1.16. Phase and gain plot of a folded cascode OTA.

1.4 Sample-and-hold amplifier

1.4.1 Introduction

In many classical SHAs, open loop architecture can achieve a high sampling rate at the cost of resolution. Operational amplifiers or operational transconductance amplifiers, when operated in closed loop topology, improve the resolution of SHAs. Other than sacrificing the accuracy, an open loop SHA also suffers from clock feed-through generated by the switch induced charge injection. Applying op-amps in a negative feedback loop can remove this feed-through to some extent [12, 13]. To further remove this error, techniques such as bottom-plate sampling can be used. Bottom-plate sampling generates a constant charge injection from the switches, which can be effectively removed by using differential architecture. Also, by using boot-strapping, switch resistances can be made independent of the signal variation, thus improving the linearity of the circuit.

1.4.2 Noise issues in S&H circuits

Any S&H circuit consists of a switch and at minimum a capacitor. The MOS switch introduces noise and it needs to be studied thoroughly.

1.4.2.1 kT/C *noise*

The switch always has some finite on-resistance which generates thermal noise.

The power spectral density of thermal noise = $4kTR$ V^2 Hz^{-1}:

k = Boltzmann's constant;

T = absolute temperature;

R = resistance.

The noise during the sampling is the resistor noise filtered by the low-pass circuit formed by the sampling capacitor and the turn-on resistance of the CMOS switch. Integrating the resistor noise spectral density weighted by the low-pass transfer function yields the mean square noise voltage on the capacitor as follows:

$$\begin{aligned}\sigma^2 &= 4kTR\int_0^\infty \frac{1}{1+(2\pi RCf)^2}df \\ &= \frac{4kTR}{(2\pi RC)^2}[2\pi RC\ \arctan(2\pi RCf)\,]_0^\infty = \frac{kT}{C}.\end{aligned} \tag{1.15}$$

By looking at the result it becomes obvious why this noise is often referred to as kT/C noise. An interesting feature is that the noise voltage does not depend on the value of the turn-on resistance of the switch, and thus the only parameter which can be used to control the noise is the size of the sampling capacitor. The desired signal bandwidth is typically at least an order of magnitude lower than the noise bandwidth of the sampling circuit; however, the sampled noise is still determined by equation (1.15). This is due to the fact that the sampling operation aliases all the noise energy into the Nyquist band. In ADCs a common requirement is that thermal noise power is smaller than the power of the quantization noise, which can be shown to be $LSB^2/12$. This sets the lowest limit for the capacitor value C as follows:

$$C > \frac{12 \cdot kT}{LSB^2} = \frac{12 \cdot kT}{2^{-2N} \cdot V_{FS}}, \tag{1.16}$$

where N is the number of bits and V_{FS} is the voltage corresponding to the ADC full-scale reading. According to equation (1.16), in the case of 1 V full-scale voltage, the capacitor values required for 10 and 16 bit resolution are 0.052 pF and 210 pF, respectively, which indicates that the capacitor values for 16 bit resolution would be too large for practical integration. To overcome this, a popular solution in high resolution applications is to use an oversampling ADC architecture, in which the capacitor size can be reduced linearly with the oversampling ratio.

1.4.2.2 Jitter in the sampling clock

Random variation of the sampling period is known as jitter. It originates from clock generator phase noise and sampling circuit noise. How the jitter is transformed to the amplitude error in the sampled voltages can be understood as follows: the error in the sampled voltage is equal to the change in the input voltage between the ideal sampling instant and the actual sampling instant. The voltage change in turn is proportional to the jitter and the rate of change of the input signal, i.e. its derivative. For a sinusoidal input the derivative is the cosine function multiplied by the corner frequency, which means that the voltage error is proportional to the frequency and the amplitude of the input signal. It can be shown that the signal-to-noise ratio limited by jitter can be written as

$$\mathrm{SNR} = -20\log(2\pi f \Delta V_t), \tag{1.17}$$

where f is the frequency of the input signal and ΔV_t is the rms value of the jitter [14].

1.4.3 Flip-around SHA

In the present work, the SHA, which is to be used as a front-end sampler of an ADC, must have a resolution higher than that of the ADC. A closed loop SHA architecture, commonly used in SC circuits and referred to as flip-around S&H amplitude, is shown in figure 1.17. Instead of using the op-amp in the negative feedback loop, this circuit uses passive circuits in the feedback, thus making faster acquisition possible. This differential circuit operates in two phases. During phase (Φ_1) switches S_1 and S_3 are closed and the capacitor C_s is charged to (V_{in}^-) − V_{CM}. In the amplification phase (Φ_2), switches S_2 and S_7 are closed. In this phase op-amp works in a closed loop and the input signal is reflected to the output.

In CMOS technology, switches are generally implemented using MOS transistors. A conducting MOS switch has a finite amount of mobile charges and when it turns off, the charges distribute themselves through the drain, source and bulk terminals of the device. There are various methods to reduce this leakage charge error which is popularly known as charge injection error.

Applying dummy switches working in opposite phase is a primitive but useful idea. The errors can be further reduced by applying transmission gates. However, if the errors can be made independent of the signal variations, superior linearity can be achieved by using differential architecture.

1.4.4 Bandwidth of the SHA

As shown in figure 1.17, the SHA works in two phases. In phase Φ_1 sampling capacitance C_s samples the input signal while the OTA remains in an open loop. During phase Φ_2 the OTA is put into the closed loop and it is at this phase

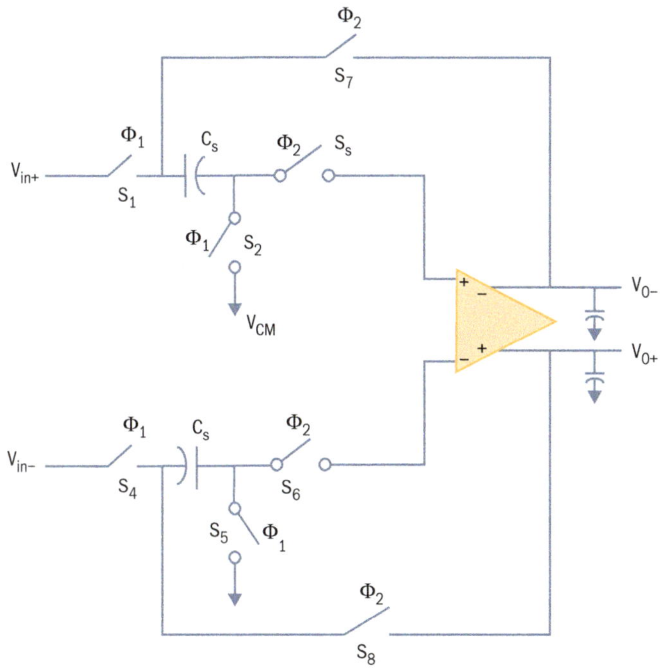

Figure 1.17. A fully differential flip-around SHA circuit.

that the SHA performs more slowly. To determine the design parameters of the SHA, calculations must be performed considering this phase Φ_2. During Φ_2 the OTA is used in closed loop mode and will have an effective load capacitance of

$$C_l = C_n + \frac{C_s C_p}{C_s + C_p}, \tag{1.18}$$

where C_n is the capacitance formed by the subsequent stage and C_p is the parasitic capacitance formed at the input node of the OTA. In closed loop operation

$$f_{-3\text{dB,closed}} = (1 + (A\beta) f_{-3\text{dB,open}}),$$

where A is the open loop gain of the OTA and β is the feedback factor. Also, from the fact that $f_{-3\text{dB,open}} = f_{u,}$ it can be deduced that

$$f_{-3\text{dB,closed}} = \beta f_u. \tag{1.19}$$

For the flip-around SHA shown in figure 1.17, the feedback factor is $\beta = \frac{C_s}{C_s + C_p}$. For a single-stage OTA, f_u is related to its transconductance g_m of OTA as $f_u = \frac{g_m}{2\pi C_l}$. Thus closed, the loop bandwidth of the SHA can be determined using the following equation

Table 1.2. Simulation result for sample and hold circuit.

Acquisition time	1.67 ns
Hold settling time	2.1 ns
Sampling speed	125 MSPS
Dynamic range	300 mV

$$f_{-3\text{dB,closed}} = \frac{1}{2\pi}\frac{C_s}{C_s + C_p}X\frac{g_m}{C_n + \frac{C_sC_p}{C_s + C_p}}. \tag{1.20}$$

Assuming $C_s \gg C_p$ and $C_n \gg C_p$, equation (1.20) reduces to

$$f_{-3\text{dB,closed}} = \frac{g_m}{2\pi C_n}. \tag{1.21}$$

For a flip-around SHA, with $C_s \gg C_p$, feedback factor β becomes unity. Thus the capacitance of the next stage will be bounded by the equation

$$C_n \leqslant \frac{g_m}{2\pi f_u}. \tag{1.22}$$

1.4.5 Results

Table 1.2 presents simulation results for the SHA circuit. Here acquisition time represents the time required for the SHA to reach the present sample magnitude from a previous sampled value. Settling time represents the time required for the SHA to settle on a final value. The transient response of the SHA circuit is shown in figure 1.18.

1.5 Multiplying digital-to-analog converter (MDAC)

1.5.1 Introduction

In a pipelined ADC, an MDAC is a critical block performing multiple jobs such as signal subtraction, multiplication and the S&H function. In this section, a detailed discussion of an MDAC to be used in an ADC is provided. The switched-capacitor circuit is used to build the MDAC for better linearity and to have reduced power consumption.

1.5.2 Implementation and working principles

Like most SC circuits, SC MDACs [15] also work with two non-overlapping clock phases. Figure 1.19 shows a typical SC MDAC circuit. A single-ended topology has been considered to explain the function of this MDAC. Subsequent small signal analysis is performed using an inverting operational amplifier. The input signal is V_{in}

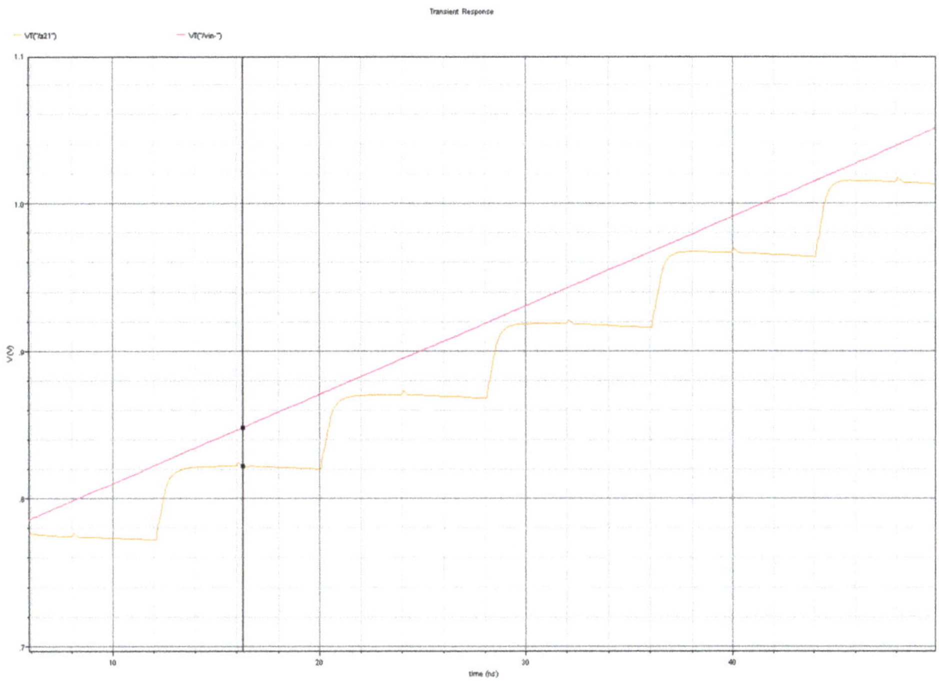

Figure 1.18. Transient behavior of S&H.

while the output signal is V_o. In clock phase Φ_1 input voltage V_{in} is charged across all the capacitors. The charge Q_1 accumulated on the capacitors can be expressed as

$$Q_1 = V_{\text{in}}(C + C + 2C + 2^2C + \cdots + 2^{N-1}C) \tag{1.23}$$

$$= V_{\text{in}}2^N C. \tag{1.24}$$

In clock phase Φ_2 the amplifier is put into a closed loop. Apart from capacitor C_i, all the other capacitors become connected to V_{ref} as defined by the sub-ADC outputs $(b_{N-1}, b_{N-2} \dots b_1, b_0)$. Assuming that the gain of the amplifier (A) is finite, charge Q_2 on the capacitors during Φ_2 is formulated as

$$\begin{aligned} Q_2 = {} & 2C\left(V_0 + \frac{V_0}{A}\right) + C \cdot V_{\text{ref}}(b_{N-1}2^{N-1} + b_{N-2}2^{N-2} + \cdots + b_1 2 + b_0) \\ & + \frac{V_0}{A}(2^{N-1} + 2^{N-2} + \cdots + 2 + 1)C. \end{aligned} \tag{1.25}$$

Assuming there is no loss of charge on the capacitor during switching of the two phases of the clock, it can be written as $Q_1 = Q_2$:

$$V_{\text{in}}2^N C = 2C\left(V_0 + \frac{V_0}{A}\right) + C \cdot V_{\text{ref}}\sum_{i=0}^{N-1} b_i 2^i + \frac{V_0}{A}(2^N - 1)C. \tag{1.26}$$

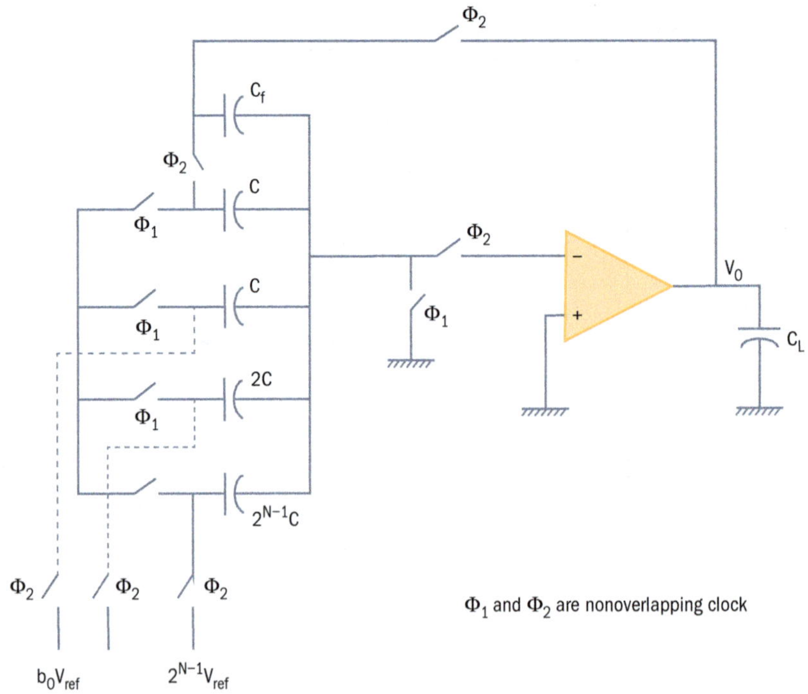

Figure 1.19. Basic block diagram of an SC MDAC circuit.

Letting $C_T = C + 2C + 2^2C + \cdots + 2^{N-1}C$, where $C_T =$ is the total capacitance:

$$V_0 = \frac{V_{\text{in}}2^{N-1} - \frac{V_{\text{ref}}}{2}\sum_{i=0}^{N-1} b_i 2^i}{\left(1 + \frac{C_T}{2AC}\right)}. \tag{1.27}$$

In order to obtain a resolution of 1.5 bits per stage it is necessary to achieve a gain of 2 in a closed loop. From the above equation, it is evident that a gain of 2 will be achieved when $N = 2$. Assuming that the open loop gain of the amplifier is very high ($A \to \infty$) and $N = 2$,

$$V_0 = 2V_{\text{in}} - (2b_1 + b_0)\frac{V_{\text{ref}}}{2}, \tag{1.28}$$

where b_1 and b_0 are the sub-ADC output. For a single-ended circuit (b_1, b_0) will lie in the range (00, 01, 10) since digital redundancy is considered. In figure 1.20 an MDAC implemented using equation (1.28) is illustrated.

Equations involving the transfer function of a 1.5 bit MDAC can be described for a single-ended circuit, with $V_{\text{ref}} = FS$,

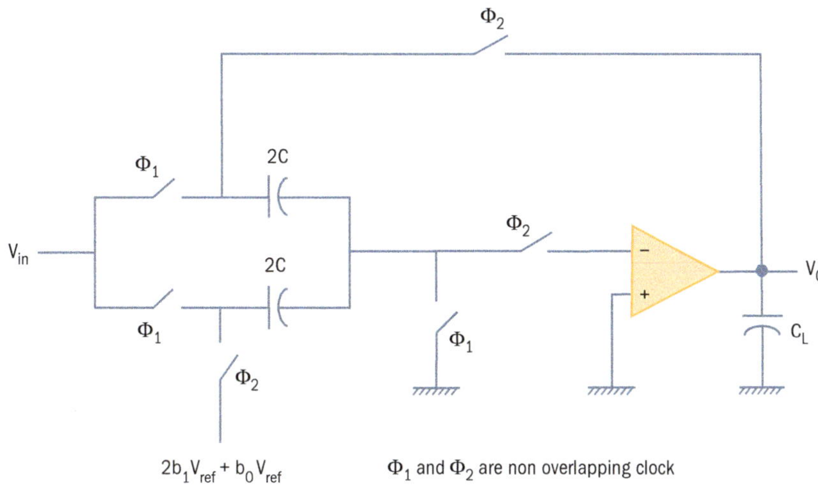

Figure 1.20. Fundamental block diagram of an SC MDAC circuit.

$$V_0 = \begin{bmatrix} 2V_{\text{in}} & V_{\text{in}} \leqslant \dfrac{3V_{\text{ref}}}{8} \\ 2V_{\text{in}} - \dfrac{V_{\text{ref}}}{2} & \dfrac{3V_{\text{ref}}}{8} \leqslant V_{\text{in}} \leqslant \dfrac{5V_{\text{ref}}}{8} \\ 2V_{\text{in}} - V_{\text{ref}} & V_{\text{in}} \geqslant \dfrac{5V_{\text{ref}}}{8} \end{bmatrix}. \tag{1.29}$$

1.5.3 Bandwidth limitations of an MDAC

The size of the sampling capacitors in an MDAC can seriously limit the sampling speed of both the MDAC and the ADC. To analyze its effect, the OTA used in the MDAC will be modeled as a single-pole device. As shown in figure 1.21, the MDAC of the first stage is in the evaluation phase, whereas that of the next stage is in the sampling phase. The load capacitance C_l of the first stage can be determined by taking the sampling capacitors of the second stage into consideration. The parasitic capacitance of the OTA, $C_{\text{OUT1},P}$ of the first stage at the output node will also contribute to its load capacitance. For simplicity, the parasitic capacitances will be ignored while analyzing the settling behavior. Load capacitance $C_{l,1}$ turns out to be

$$C_{l,1} = 2C_1 + 2^{n_2}C_2, \tag{1.30}$$

where C_1 is the sampling capacitor of the first stage while C_2 is that of the second stage. n_1 and n_2 are the stage resolutions of the first and second stage, respectively. As mentioned before, the OTA discussed is single-pole, so the settling error occurring from this OTA would be

$$\varepsilon_r = e^{-t_s\omega_b}, \tag{1.31}$$

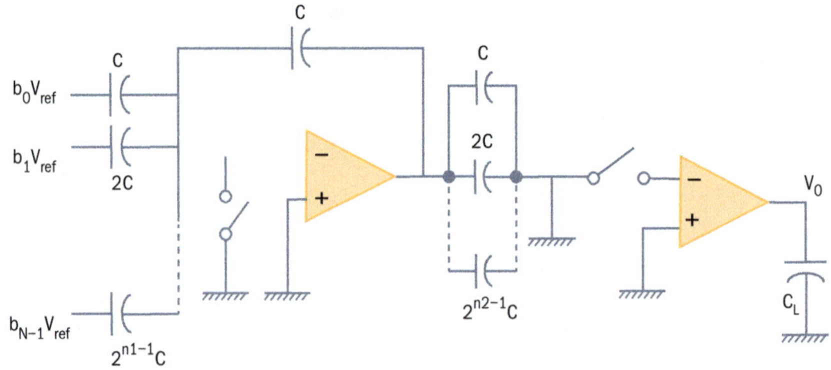

Figure 1.21. Capacitive loading of an MDAC.

where t_s is the settling time and ω_b is the 3 dB bandwidth of the OTA. Since the OTA is in a closed loop during the sampling mode, the feedback factor β can be calculated as

$$\beta = \frac{2}{2^{n_1} + 1}. \tag{1.32}$$

Also, as the 3 dB frequency ω_b and unity gain frequency (UGB) ω_u of the OTA are related as $\omega_b = \omega_u\beta$, it can be further rewritten as

$$\omega_b = \frac{g_{m1}}{C_{l,1}}\beta \tag{1.33}$$

$$\omega_b = \frac{2g_{m_1}}{2^{n_2}C_2(2^{n_1} + 1) + 2C_1(2^{n_1} + 1)}. \tag{1.34}$$

Since an OTA is essentially a single-stage amplifier, the load capacitance will define its UGB, ω_u. Thus, the equation $\omega_u = \frac{g_m}{C_l}$ will hold, where g_m is the transconductance of the OTA. Also, MDAC must work fast enough so that the output of the first stage settles to $\frac{\text{LSB}}{2}$, where LSB would be as determined from the second stage onwards, i.e. $\text{LSB} = \frac{1}{2^{n_{\text{TOT}}-n_1+1}}$. Here, it has been assumed that digital error correction is applied to the first stage. Also, considering that the MDAC must settle within half of the sampling period, $t_s = \frac{1}{2f_s}$,

$$\varepsilon_r = \mathrm{e}^{-t_s w_{-3\text{dB}}} \leqslant \frac{1}{2^{n_{\text{TOT}}-n_1+2}}. \tag{1.35}$$

From the thermal noise consideration equation (1.16),

$$C_2 \geqslant \frac{C_1}{2^{n_2-1}}. \tag{1.36}$$

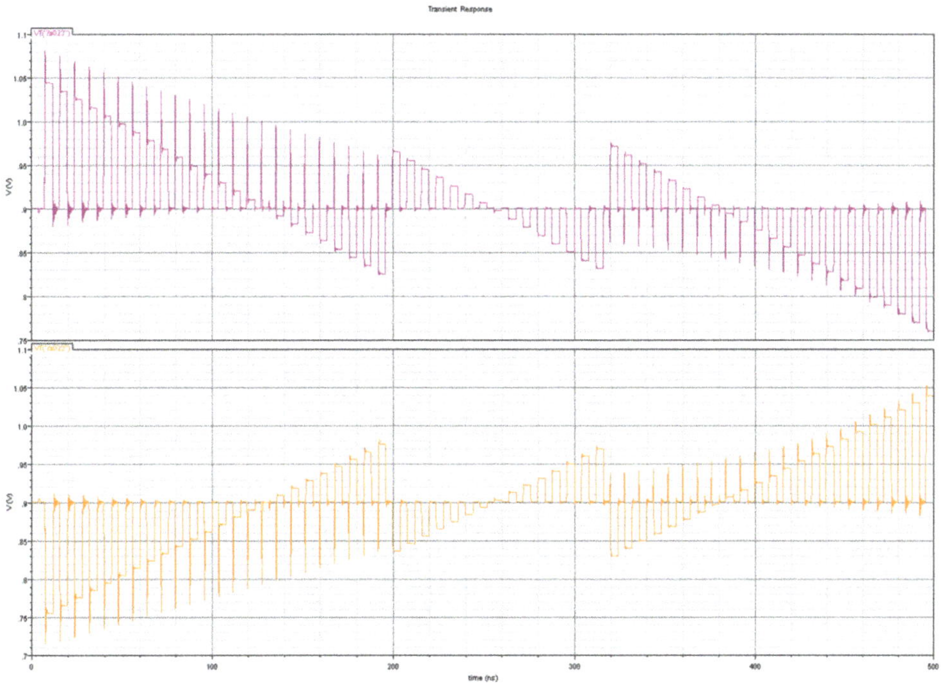

Figure 1.22. Ramp response of an MDAC.

From the above equations (1.34)–(1.36)

$$C_1 \leqslant \frac{g_{m1}}{4f_s(2^{n1}+1)\ln(2^{n_{\mathrm{TOT}}-n_1+2})}. \tag{1.37}$$

Thus, the unit sampling capacitors of the first stage and the consecutive stages can be measured from equations (1.6), (1.7), (1.36) and (1.37). These equations can be summarized as follows:

$$\frac{3.2^{2n_{\mathrm{TOT}}+2}kT}{2^{n_1}\,\mathrm{FS}^2} \leqslant C_1 \leqslant \frac{g_{m1}}{4f_s(2^{n1}+1)\ln(2^{n_{\mathrm{TOT}}-n_1+2})} \tag{1.38}$$

$$\frac{C_1}{2^{n_2-1}} \leqslant C_2 \leqslant 2^{n_1-n_2}C_1. \tag{1.39}$$

1.5.4 Results

Figure 1.22 shows the MDAC output for a positive and negative slope ramp signal with magnitude 300 mV. Here the sampling frequency is 125 MSPS.

1.6 Comparator

1.6.1 Introduction

Comparators are the major constituent blocks of sub-ADCs in pipelined ADCs. Along with the OTA, the speed of the comparator also limits the sampling speed of the ADC.

For proper operation, the comparator output must settle to its final value within the clock time period T_{clk}. A static comparator 0 inherently has a very high open loop gain, and thus shows very high accuracy. On the other hand, a dynamic comparator works under the control of the clock signal, and high speed and very low power consumption are possible using these clocked circuits.

In this chapter, a static comparator has been improvised to transform it into a dynamic comparator. Usually a static comparator consists of a high gain pre-amplifier, a high gain latch and finally a buffer to drive capacitor loads. The high gain pre-amplifier is primarily responsible for the overall high open loop gain of the comparator. This high gain also reduces the input referred offset voltage, thus improving the accuracy of the amplification circuit, which, in this case, is the sub-ADC of a pipelined ADC. These static comparators are often used in switch capacitor circuits with MOS switches at its input. Charge injection error and clock feed-through caused by these switches distorts the analog signal, thus degrading the switch capacitor circuit performance. To alleviate such complexities in the present design, the latch block of a static comparator has been clocked to perform the switching without compromising the circuit accuracy. Further, dynamic comparators usually suffer from kick back noise, which occurs mainly due to its low open loop gain and clock transitions. However, high pre-amplifier gain of the designed comparator helps suppress the possible kick back noise which might arise due to fast clocking.

1.6.2 Dynamic comparator

The dynamic comparator consists of four major parts at the front: a high gain pre-amplifier, a latch, a self-biased differential amplifier and an output driver for driving the large capacitive loads.

A pre-amplifier is basically a differential amplifier with a diode connected load, as shown in figure 1.23. The two transistors M_5 and M_6 determine the bias currents of the two differential pairs M_1–M_2 and M_3–M_4, respectively. The threshold voltage of the comparator is determined by the current division in the differential pairs and between the cross-coupled branches.

1.6.3 Input pre-amplifier stage

The determination of the switching point of the comparator can be modeled with the simplification of figure 1.24 [15] for the two cross-coupled differential pairs. Using the symbols indicated in the figure and having $W_1 = W_2$, $W_3 = W_4$, the transistors M_1–M_4 follow the large signal current equations as follows:

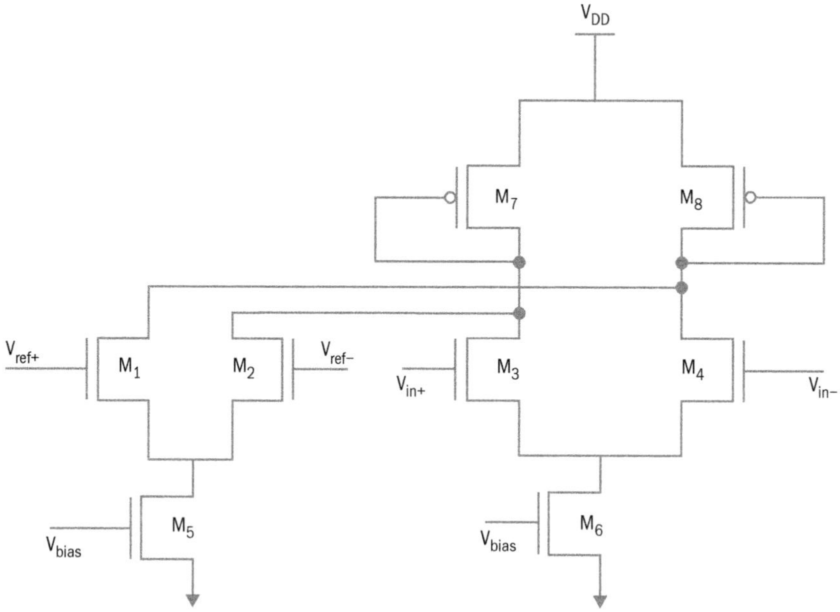

Figure 1.23. Pre-amplifier stage for the dynamic comparator.

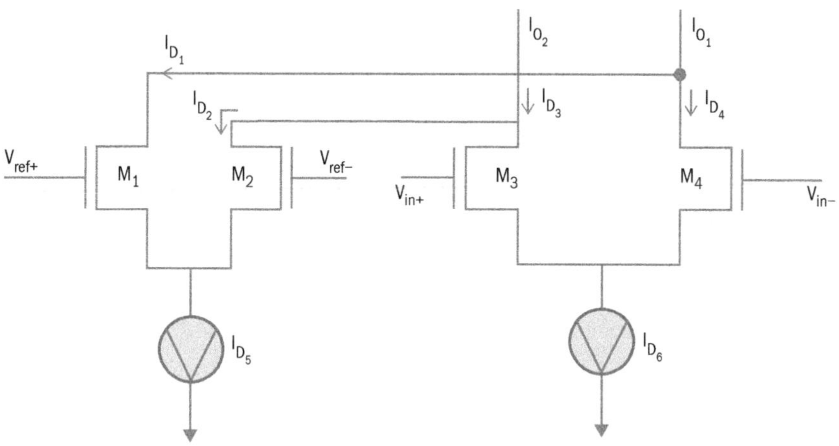

Figure 1.24. Simplified model of the differential pair comparator.

$$I_{D_1} - I_{D_2} = \beta_1 \cdot V_{\text{in}} \cdot \sqrt{\frac{2I_{D_5}}{\beta_1} - V_{\text{in}}^2} \tag{1.40}$$

$$I_{D_3} - I_{D_4} = \beta_3 \cdot V_{\text{in}} \cdot \sqrt{\frac{2I_{D_6}}{\beta_3} - V_{\text{ref}}^2} \tag{1.41}$$

$$\beta_i = \mu_o C_{\text{ox}} \frac{W_i}{L_i}$$
$$V_{\text{in}} = V_{\text{in}}^{+} - V_{\text{in}}^{-}$$
$$V_{\text{ref}} = V_{\text{ref}}^{+} - V_{\text{ref}}^{-}.$$

The comparator changes its outputs when the currents I_{o1} and I_{o2} of both output branches are equal. Assuming the relation of the source coupled pair bias currents to be $I_{d_3} = d \cdot I_{d_8}$ and by marking the threshold point with parameter e so that $V_{\text{in}} = e \cdot V_{\text{ref}}$, this results in a condition

$$2 \cdot d \cdot e \cdot I_{d_3} \frac{W_1}{L} - \mu_o C_{\text{ox}} e^4 V_{\text{ref}}^2 \left(\frac{W_1}{L}\right)^2 = I_{d_6} \frac{W_3}{L} - \mu_o C_{\text{ox}} V_{\text{ref}}^2 \left(\frac{W_3}{L}\right)^2. \tag{1.42}$$

When the parameters d and e are chosen according to the wanted threshold point of the comparator, the transistor dimensions W_1 and W_3 can be interpolated from equation (1.42).

1.6.4 The latch

Regenerative comparators use positive feedback to accomplish the comparison of two signals. The regenerative comparator is also called a latch. The simplest form of a latch [10] is shown in figure 1.25. It consists of two cross-coupled NMOSs. The current sources are used to identify the dc currents in the transistor. The latch has two modes of operation. The first mode disables the positive feedback and applies the input signal to the terminals designated as v_{o_1} and v_{o_2}.

The initial voltage applied during this mode will be designated as v_{o_1}' and v_{o_2}'. The second mode enables the latch and, depending upon the relative values of v_{o_1}' and v_{o_2}', one of the outputs will go high and the other will go low. A two-phase clock is used to determine the mode of operation.

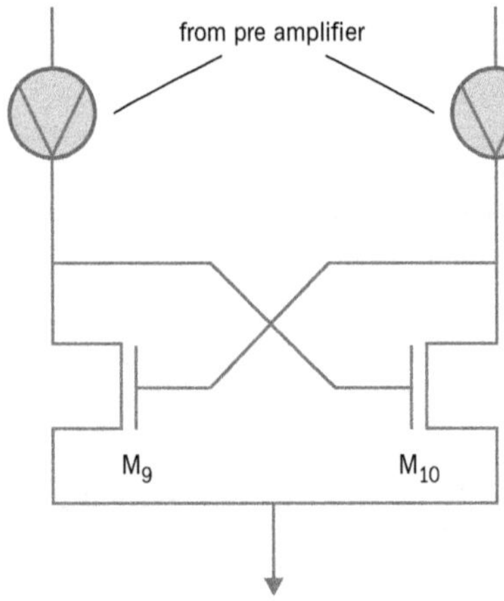

Figure 1.25. NMOS latch.

The output difference voltage for a given input difference is

$$\nabla v_o(t) = e^{\frac{g_m R}{\tau}t} \cdot \nabla v_i, \tag{1.43}$$

$\nabla v_o(t)$ = output difference voltage at v_{o_1} and v_{o_2} terminals after positive feedback enabled;
∇v_i = input difference voltage at at v_{o_1} and v_{o_2} terminals before positive feedback enabled;
g_m = transconductance of MOSFET;
R = output resistance of MOSFET;
τ = time constant of latch (RC output capacitance at v_{o_1} or v_{o_2}).

From the above equation it is clear that the output response of the system is directly proportional to the magnitude of the input voltage difference.

1.6.5 Self-biased differential amplifier or complimentary self-differential amplifier (CSDA)

The self-biased differential amplifier [16–19] is illustrated in figure 1.26, in which M_3 and M_4 are biased with internal amplifier node V_{bias}. This self-biasing of the amplifier creates a negative feedback loop that stabilizes the bias voltages. Any variations in processing parameters or operating conditions that shift the bias voltages away from their nominal values result in a shift in V_{bias} that corrects the

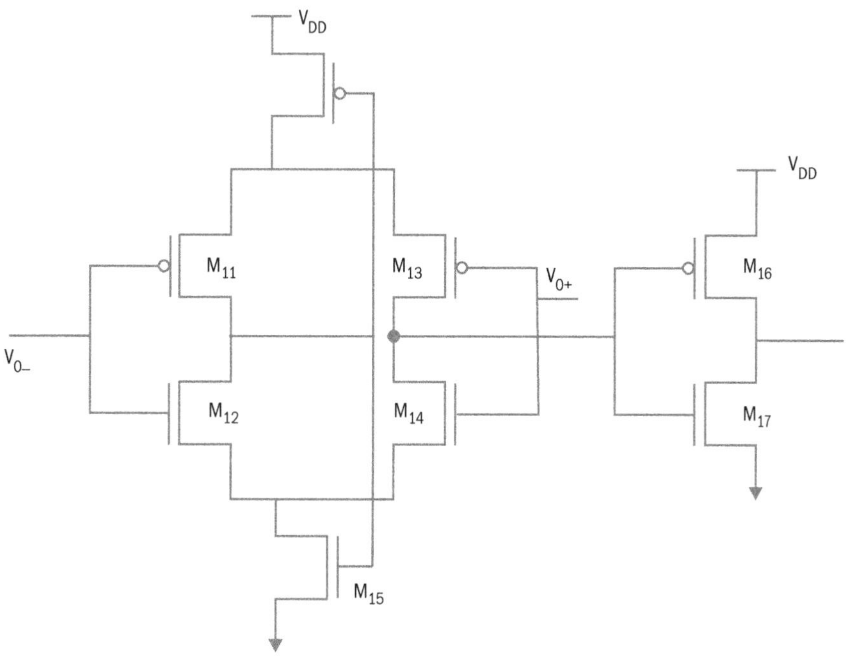

Figure 1.26. Self-biased differential amplifier.

bias voltages through negative feedback. In the CSDA, devices M_3 and M_4 operate in the linear region. Consequently, the voltages V_H and V_L may be set very close to the supply voltages. Since these two voltages determine the output swing of the amplifier, the output swing can be very close to the difference between the two supply rails. This large output swing makes interfacing the CSDA to ordinary CMOS logic gates straightforward, since it provides a large margin for variations in the logic threshold of the gates.

Another consequence of the linear-region operation of devices M_3 and M_4 is that the CSDA can provide output switching currents that are significantly greater than its quiescent current. In contrast, conventional CMOS differential amplifiers cannot provide switching currents that exceed the quiescent current set by the current-source device, which operates in the saturation region. This capability of supplying momentarily large current pulses makes the CSDA particularly suitable for high speed comparator applications, where it is necessary to rapidly charge and discharge output capacitive loads without at the same time consuming inordinate amounts of power. The complementary character of the CSDA affords it an approximate doubling (+6 dB) in dc differential-mode gain over conventional amplifiers. The differential-mode gain A_d of the CSDA is given by

$$A_d = \frac{g_{m1} + g_{m2}}{g_o}, \tag{1.44}$$

where g_{m1} and g_{m2} are the transconductances of devices $M1_{a-b}$ and $M2_{a-b}$, respectively, and g_o is the output conductance of the amplifier.

The complete schematic view of the comparator is shown in figure 1.27. At the front end a high gain pre-amplifier provides the major components of the overall open loop gain of the comparator. Next, a high gain latch also adds to the gain of

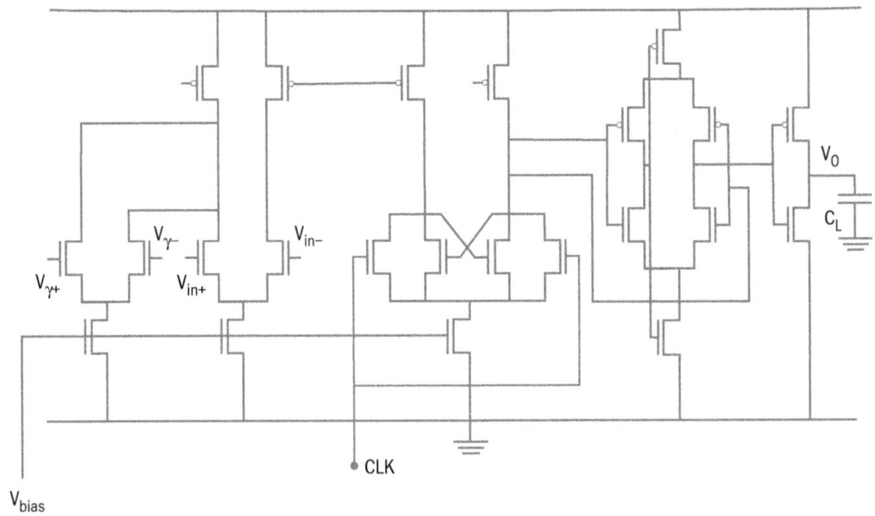

Figure 1.27. Circuit schematic of a dynamic comparator.

the circuit. This latch is clocked and this is where it differs from the standard static comparator. The latch is followed by a self-biased differential amplifier which converts the differential signal to a single-ended signal. This signal is then passed through a buffer which gives it driving capability for large capacitive loads. The operation of the comparator can be simplified with the help of the following equation:

$$\begin{aligned} V_{\text{out}} &= A_{\text{open}} \{(V_{\text{ref}+} - V_{\text{ref}-}) - (V_{\text{in}+} - V_{\text{in}-})\} \text{ if } V_{\text{clk}} = 0 \\ V_{\text{out}} &= V_{\text{DD}} \text{ if } (V_{\text{ref}+} - V_{\text{ref}-}) > (V_{\text{in}+} - V_{\text{in}-}),\ V_{\text{clk}} = 0 \\ &= 0 \text{ if } (V_{\text{ref}+} - V_{\text{ref}-}) \leqslant (V_{\text{in}+} - V_{\text{in}-}),\ V_{\text{clk}} = 0 \\ &= 0 \text{ if } V_{\text{clk}} = V_{\text{DD}}. \end{aligned} \tag{1.45}$$

1.6.6 Results

Transient and frequency domain simulations of the designed dynamic comparator are listed in table 1.3 and the response to the sinusoidal input is as shown in figure 1.28.

1.7 Conclusion

1.7.1 Summary

This chapter mainly concentrates on the implementation and design blocks of a 10 bit, 500 MSPS parallel pipelined analog-to-digital converter. The required blocks were implemented using 180 nm standard CMOS technology.

The proposed pipelined ADC uses analog preprocessing to divide the input signal range into sub-intervals and amplification of a residue signal for further processing in the subsequent stages. Each stage gives two bits, one bit for digital output and the other for digital error correction. The realizations of the preprocessing stages have been implemented using switched-capacitor circuits. In this chapter two major bocks were implemented using switched-capacitor techniques—the front end sample-and-hold amplifier (SHA) and the multiplying digital-to-analog converter (MDAC). Both these circuits are implemented by amplifying the device at its core. In this work, the folded cascode OTA, SHA, MDAC and dynamic comparator are described along with their design details and simulation results.

Table 1.3. Performance parameters.

Performance	Simulation results
Input offset (resolution) voltage	5 mV
Input dynamic range	900 mV peak to peak
Power consumption	1.06 mW
Input clock frequency	⩽ 300 MSPS

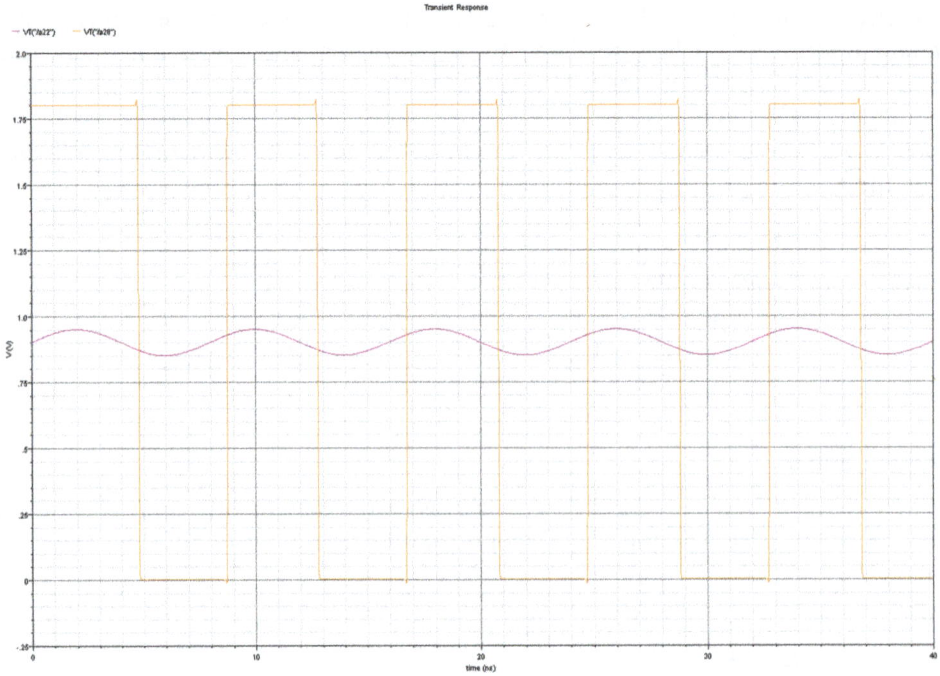

Figure 1.28. Dynamic comparator output for sinusoidal input.

1.7.2 Future work

In this work most of the analog blocks were designed which are required for a parallel pipelined ADC. The complete digital output of the ADC is the combination of all the bits from each stage. For a given input the digital output from each stage is delayed by $T_s/2$ and these bits have to be added according to the RSD algorithm. To achieve this outcome, the bit alignment and digital-correction block has to be implemented.

The major disadvantage of the parallel pipelined ADC is its huge power consumption. In pipelined ADC, during the sampling mode OTA is in the idle stage. If we share this idle time with other channels for amplification, the number of OTAs will be drastically reduced and thereby the power consumption will be reduced. To achieve this the amplifier sharing technique has to be implemented.

References

[1] Walt K 2004 *Analog to Digital Converters* (Philippines: Analog Devices Inc.)

[2] Cho T B 1995 Low-power low-voltage analog-to-digital conversion techniques using pipelined architectures *Memorandum No. UCB/ERL M95/23* (Berkeley, CA: Electronics Research Laboratory)

[3] Lewis H, Fetterman H S, Gross G F, Ramachandran R and Viswanathan T R 1992 A 10-b 20-M sample/s analog-to-digital converter *IEEE J. Solid-State Circuits* **27** 351–8

[4] Genetti B and Jespers P 1990 A 1.5 MS/s 8 bit pipelined RSD A/D converter *ESS-CIRC Dig. Tech. papers* pp 137–40

[5] Zhang S, Huang L and Lin B 2007 Design of a low-power, high speed op-amp for 10 bit 300 MSPS parallel pipelined ADC *Proc. of the 2007 IEEE Int. Conf. on Integration Technology (Shenzhen, China, 20–24 March 2007)*

[6] Min B-M 2003 A 69-mW 10 bit 80MSPS pipelined CMOS ADC *IEEE J. Solid-State Circuits* **38** 2031–9

[7] E-Waltari M and Halonen K A I 2002 *Circuit Techniques for Low-voltage and High Speed A/D Converters* (Dordrecht: Kluwer)

[8] Vallee R E and El-Masry E I 1994 A very high-frequency CMOS complementary folded cascode amplifier *IEEE J. Solid-State Circuits* **29** 130–3

[9] Behzad R 2002 *Design of Analog CMOS Integrated Circuits* (New York: McGraw-Hill)

[10] Hosticka B J 1979 Improvement of the gain of MOS amplifiers *IEEE J. Solid-State Circuits* **14** 1111–4

[11] Allen P E and Holberg D R 2002 *CMOS Analog Circuit Design* (Oxford: Oxford University Press)

[12] Gray P R, Hurst P J, Lewis S H and Meyer R G 2008 *Analysis and Design of Analog Integrated Circuit* 4th edn (India: Wiley)

[13] Li J and Moon U-K 2003 Background calibration techniques for multistage pipelined ADCs with digital redundancy *IEEE Trans. Circuits Syst.—II: Analog Digit. Signal Process.* **50** 531–8

[14] Shingawa M and Akazawa Y 1990 Jitter analysis of high-speed sampling systems *IEEE J. Solid-State Circuits* **25** 220–4

[15] Dai L and Harjani R 2000 CMOS switched-op-amp-based sample-and-hold circuit *IEEE J. Solid-State Circuits* **35** 109–13

[16] Sumunen L, Wulturi M and Hulonen K 2000 *A Mismatch Insensitive CMOS Dynamic Comparator for Pipelined A/D Converters* (Helsinki: Helsinki University of Technology, Electronic Circuit Design Laboratory)

[17] Andersen T N and Hernes B 2005 A cost-efficient high-speed 12-bit pipeline ADC in 0.18-m digital CMOS *IEEE J. Solid-State Circuits* **40** 1506–13

[18] Vallee R E and El-Masry E I 1994 A very high-frequency CMOS complementary folded cascode amplifier *IEEE J. Solid-State Circuits* **29** 130–3

[19] Bazes M 1991 Two novel fully complementary self-biased CMOS differential amplifiers *IEEE J. Solid-State Circuits* **26** 165–8

Chapter 2

A built-in self-test for a 1.8 V, 8 bit, 125 mega samples per second pipelined analog-to-digital converter

Md Tausiff and Alok Barua

With the fast advancement of CMOS fabrication technology, more and more signal-processing functions are implemented in the digital domain for lower cost, lower power consumption, higher yield and higher re-configurability. This has recently generated a great demand to meet the following challenges: high linearity, high dynamic range and high sampling speed, and low spurious spectral performance simultaneously under low supply voltages in deep submicron CMOS technology with low power consumption.

This chapter addresses these challenges using the pipelined analog-to-digital converter (ADC) as a demonstration platform. The proposed pipelined ADC uses analog preprocessing to divide the input signal range into sub-intervals and amplification of a residue signal for further processing in the subsequent stages. The realization of the preprocessing stages has been implemented using switched-capacitor (SC) circuits. In designing the current pipelined ADC, two major blocks were implemented using an SC technique front-end sample-and-hold amplifier (SHA) and a multiplying digital-to-analog converter (MDAC). The CMOS process was chosen as the design platform for the current ADC, and the loads are capacitive in nature. Consequently, the operational transconductance amplifier (OTA) is preferred over op-amps as an amplifying device. In this chapter, the folded cascode OTA, SHA, MDAC and dynamic comparator are described and their design details and simulation results are also presented.

2.1 Organization of the chapter

A literature survey for built-in self-tests (BISTs) on ADCs was first performed. The chapter begins with the design of a pipelined ADC with the required specifications, which are given in the next section. An 8 bit parallel pipelined ADC is designed. In

doi:10.1088/978-0-7503-1732-0ch2

this process the blocks of the pipelined ADC are designed, which include the dynamic comparator and folded cascode OTA. Before designing the components in the Cadence SPECTRE environment, we develop a model of pipelined ADC in Simulink of MATLAB to determine the principles of operation of the pipelined ADC and the characteristics of different blocks of the pipelined ADC. From this we obtain the waveforms at different stages of the pipelined ADC, which include the residue waveform, and their digital characteristics. Subsequently the design is simulated in a Cadence environment. The dynamic comparator and folded cascode OTA are the next two blocks to be designed. The next part of the project is testing the designed pipeline ADC, which should be on-chip with its BIST system. A literature survey of the different types of BIST architectures has been performed.

2.2 Specifications of the pipelined ADC

Table 2.1 displays the required specifications and the specifications which are achieved after design.

2.3 Motivation and aims

With the explosive growth of wireless communication systems and portable devices, the power reduction of integrated circuits has become a major challenge. In applications such as personal communication systems (PCSs), cellular phones, camcorders and portable storage devices, low power dissipation and hence longer battery lifetime is the principal requirement. An example of a low power application is a wireless communication system. With the rapid growth of the Internet and information-on-demand, handheld wireless terminals are becoming increasingly popular. With limited energy in a reasonable size battery, minimum power dissipation in the integrated circuit is necessary.

With the rapid growth of the 'information superhighway', large amounts of data are stored in storage devices and are accessed frequently. In order to transmit these data in a short period of time, a high transfer rate in storage devices is required. This translates directly into a higher data conversion rate in the read channels of magnetic

Table 2.1. Target and achieved specifications of the designed pipelined ADC.

Specifications	Target	Achieved
Gain	>50 dB	69.78 dB
Unity gain bandwidth	>560 MHz	770 MHz
ICMR	300 mV	300 mV
OCMR	300 mV	300 mV
Slew rate		572 V μS^{-1}
Settling time	<4 ns	0.418 nS
CMRR	>100 dB	126.07 dB
Power consumption	As small as possible	3.26 mW

storage devices. However, in order to achieve an even higher transfer rate for some multimedia applications, the speed of the ADC needs to be improved.

To achieve the goals mentioned above (i.e. low power, low voltage and high speed), CMOS technology is very attractive for several reasons [1, 2]. First, its low cost and high integration levels have made CMOS technology superior to bipolar technology. Because of this, several low power CMOS design techniques have been developed. Also, the scaled CMOS technology can achieve the high speeds which were once reserved for bipolar or other fast processes.

With the above motivations, the goal of this research is to build a high speed, low power and low voltage ADC in 0.18 μm CMOS technology.

2.4 Pipelined ADC architecture

A pipelined ADC is inherently a multi-step amplitude quantizer in which the digitization is performed by a cascade of many topologically similar or identical stages of low-resolution ADCs. Pipelining enables high conversion throughput by inserting analog registers, i.e. SHAs, in between stages that allow a concurrent operation of all stages. This is achieved at the cost of increased latency. Moreover, pipelined ADC is a good compromise between speed and circuit complexity. A block diagram of a pipelined ADC is shown in figure 2.1. A pipelined stage takes two actions when an input signal arrives (sampled by a master clock): output from the sample-and-hold (S&H) circuit and a coarse quantized signal from the sub-ADC. These two operations are often performed simultaneously or in tandem. The resolution of a typical pipelined stage is typically no more than four bits. The resolution of the conversion is enhanced by passing a residue signal—the unconverted part of the input signal—to the later stages where it is further quantized

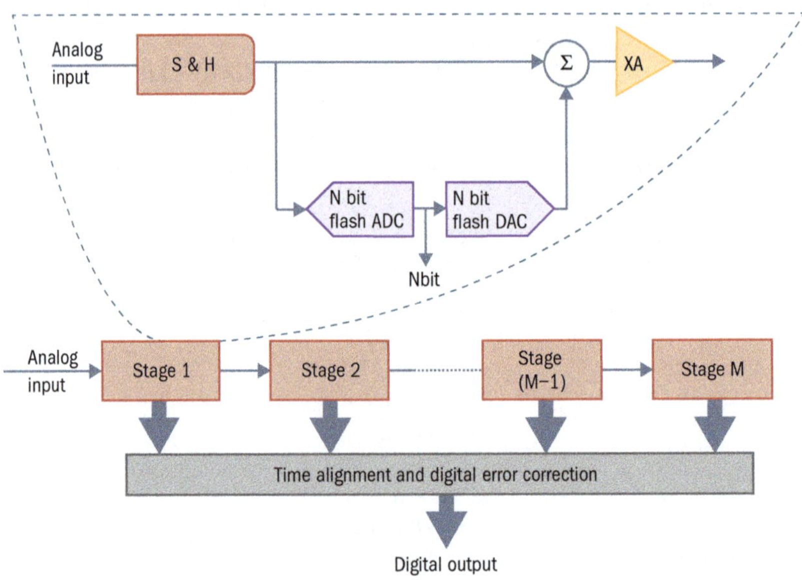

Figure 2.1. Block diagram of a pipelined ADC.

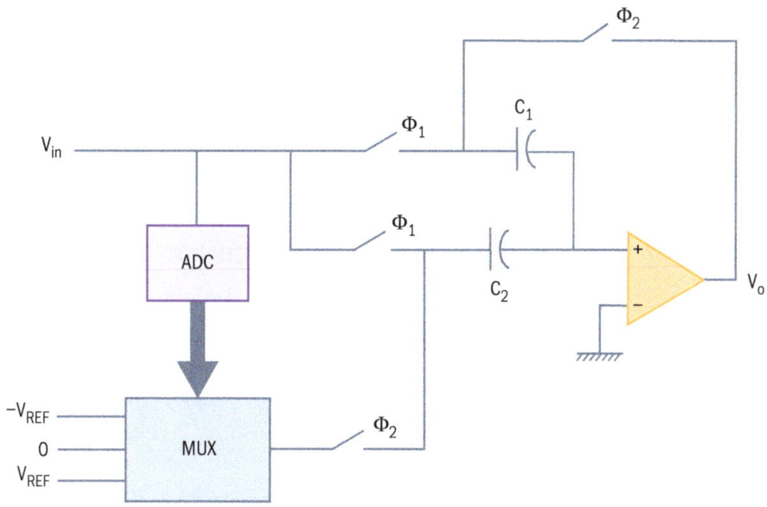

Figure 2.2. Circuit diagram for a 1.5 bits s^{-1} MDAC.

(figure 2.2). φ_1 and φ_2 are the non-overlapping 50% duty cycle clock. The conversion residue is created by a digital-to-analog converter (DAC) and a subtraction circuit. The maximum swing of this residue signal is often brought back to the full-scale reference level with a precision amplifier—the residue amplifier. This keeps the signal level constant and allows the sharing of an identical reference throughout the pipelined stages.

Dividing a high resolution conversion process into multiple steps greatly reduces the total number of comparators required in contrast to a flash converter. In the limiting case, a 1 bit/stage (bits/s) pipelined ADC only needs N comparators to resolve an N bit code as opposed to the $(2^N - 1)$ comparators required by a flash converter. The large accumulative inter-stage gain also reduces the impact of circuit non-idealities, such as noise, non-linearity and offset of later stages, on the overall conversion accuracy. For medium-to-high resolution Nyquist applications, pipelined ADCs have been demonstrated to achieve the lowest power consumption at relatively high conversion rates. In CMOS circuit technology, a typical pipelined ADC stage usually consists of a coarse comparator and a compact SC circuit termed the multiplier DAC (MDAC), which incorporates the S&H circuit, the DAC, the subtraction and the residue-gain functions. The circuit diagram of a single-ended 1.5 bits s^{-1} MDAC is shown in figure 2.2. This architecture is also known to tolerate large comparator offsets due to the built-in decision level that overlaps between successive stages, usually referred to as digital redundancy or digital error correction. The conversion accuracy thus relies solely on the precision of the residue signals; the conversion speed, on the other hand, is largely determined by the settling time of the residue amplifier [3–8].

2.5 A MATLAB model of the pipelined ADC

The model of the pipelined ADC is designed and simulated in Simulink of MATLAB to understand the concepts of the pipelined ADC and the following results are obtained. The MATLAB model is shown in figure 2.3. The simulation outputs are shown in figures 2.4–2.8.

2.6 Results obtained in the Cadence environment

It is clear from figures 2.4–2.7 that the output of the pipelined ADC which is obtained using Cadence resembles the same obtained using MATLAB.

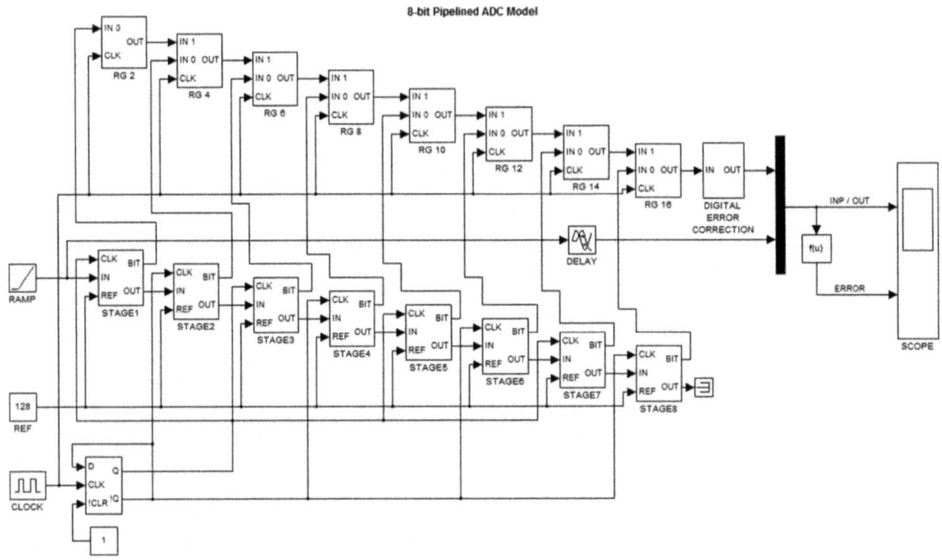

Figure 2.3. MATLAB model of the pipelined ADC.

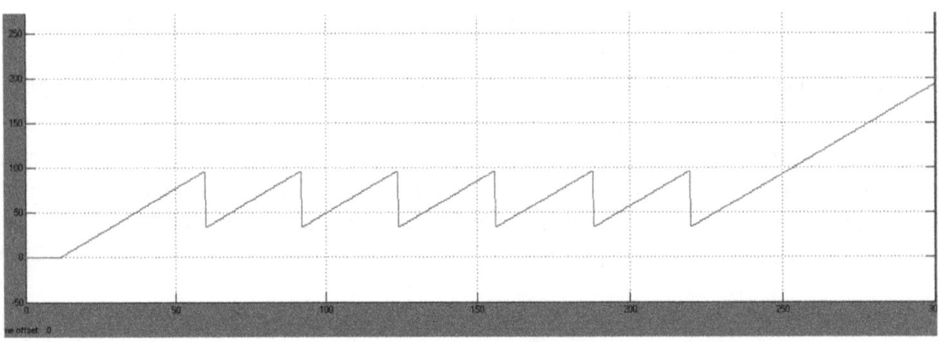

Figure 2.4. Result obtained at the output of the MDAC using MATLAB.

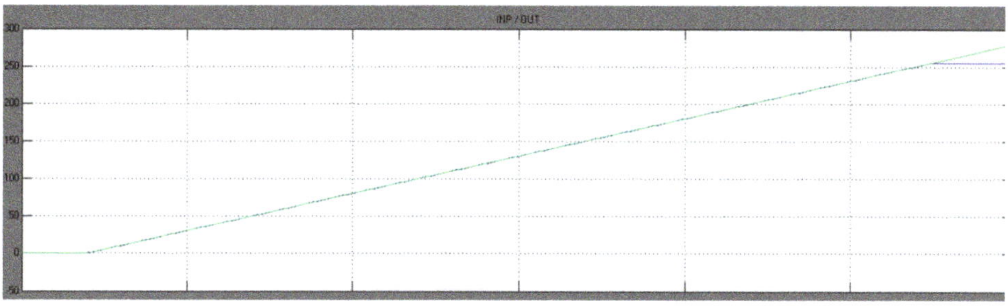

Figure 2.5. Final output of the pipelined ADC in the MATLAB model.

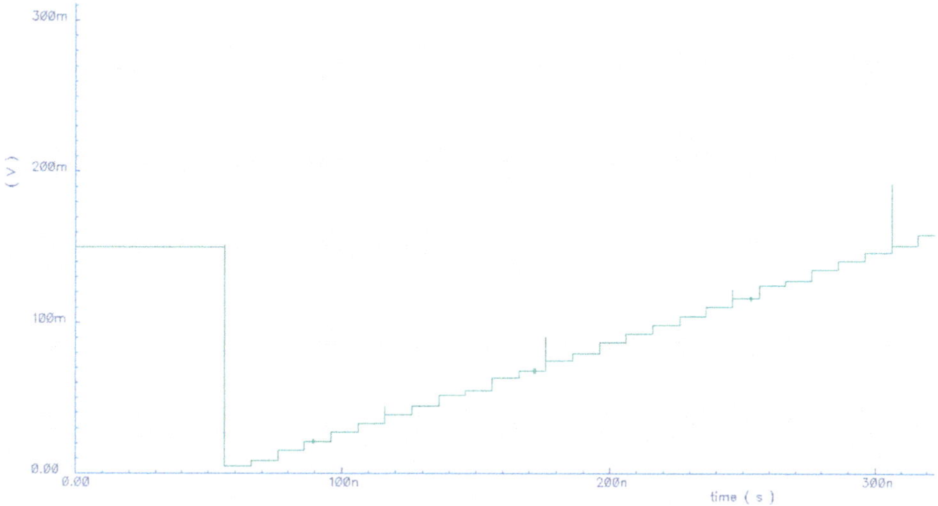

Figure 2.6. Final output of the pipelined ADC using Cadence.

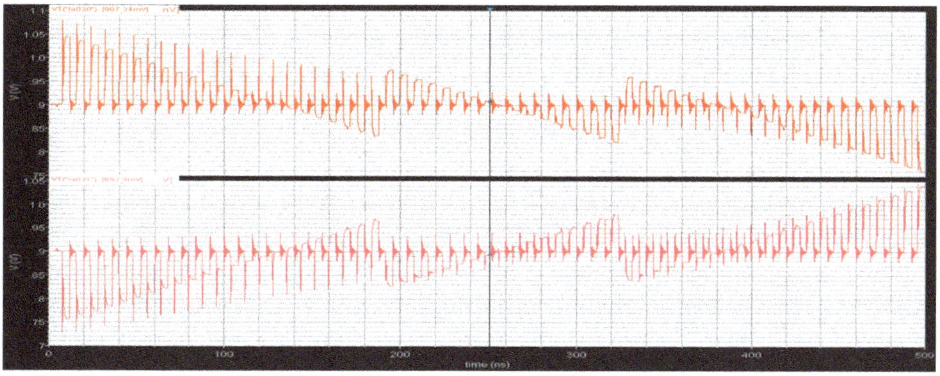

Figure 2.7. Plot of the MDAC output using Cadence.

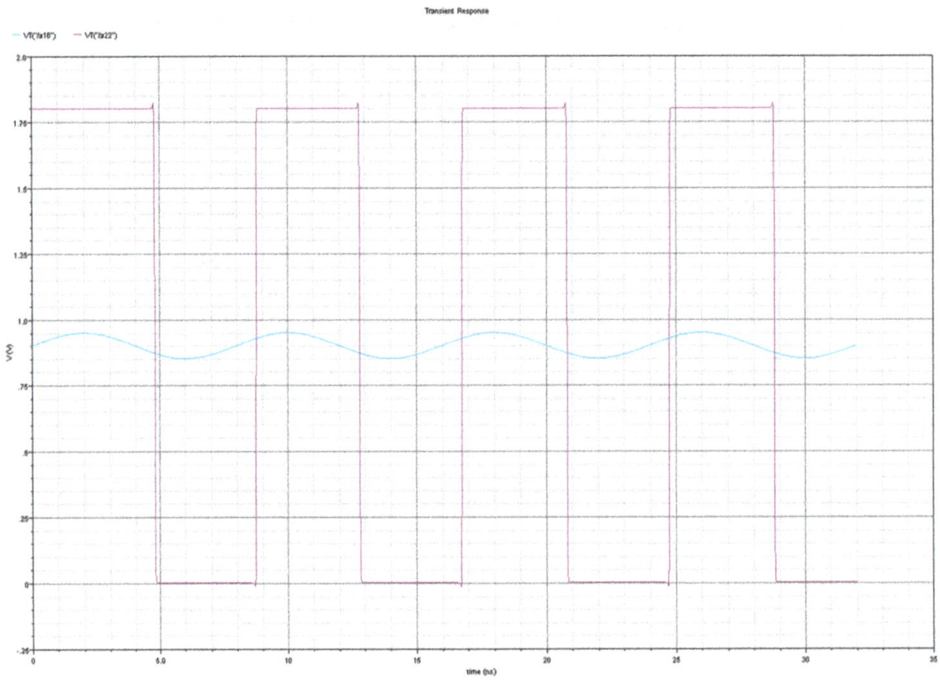

Figure 2.8. Dynamic comparator output.

2.7 Built-in self-test (BIST) system

As IC fabrication techniques have improved, a complex system can now be implemented on a chip. The testing costs of a system-on-chip (SOC) often overwhelm the design and fabrication costs due to the complexity of the implemented system. Because the price of the automatic test equipment drastically increases as the system speed and complexity increases, various BIST techniques have been introduced to reduce the test costs. Most of the reported BIST schemes are limited to digital systems and semiconductor memory. Few analog BIST schemes have been reported and this field is still in its preliminary stages. The absence of compact circuits to measure analog values at different nodes and the fuzziness of the criteria make the development of an analog BIST a challenging task [9].

ADCs and DACs are commonly included in an SOC. The ADC/DAC BIST is becoming a bottleneck for SOC testing. Most of the reported ADC/DAC schemes require a large overhead while the efficiency is low. Common ADC testing involves measuring the effective resolution and histogram with a sinusoidal test input signal. This dynamic testing requires a digital signal processor (DSP) for calculating the fast Fourier transform (FFT) or accumulators and memories which require a large overhead. The measurement of the integral non-linearity (INL) error and the differential non-linearity (DNL) error with a ramp test signal is the other ADC testing method which is suitable for BISTs due to the small overhead.

The ADC is an analog block, although the output of the ADC is digital. This property provides the possibility to implement an ADC BIST scheme with a compact digital circuit.

2.7.1 A BIST scheme for an ADC

The proposed BIST system is depicted in figure 2.9. As the analog test signal is applied to the ADC's input, the ADC generates a digital code corresponding to the input signal. Only with the digital code does the error detector decide whether the ADC has any faults or not. The ramp signal generator should have higher linearity than the ADC under test. A high linearity ramp signal generator using an SC circuit was reported in [10]. The ramp test signal can be supplied by external equipment and can be shared by many ADCs under testing in parallel. The ramp test signal varies within the dynamic range of the ADC for proper operation and should be increased by one least significant bit (LSB) per clock. Since the analog input is increased by 1 LSB per clock, the ADC output should be increased by 1 LSB per sample. If there is any fault in the system then the ADC output may not increase by 1 LSB per sample.

Four types of errors are defined in table 2.2 to detect faulty circuits. Here $x(n)$ is the ADC output code for the nth sample. If the magnitude of the difference between two consecutive samples, $x(n) - x(n-1)$, is greater than 1 LSB then this means that the ADC output increases more than 1 LSB while the input signal is increased by 1 LSB. The ADC can be considered to be faulty, namely presenting a missing code error. If the sign of the difference is negative, this means that the ADC output

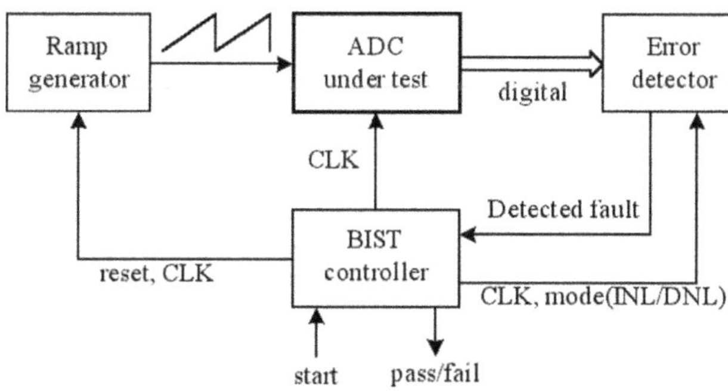

Figure 2.9. Block diagram of the BIST architecture.

Table 2.2. Classification of error types.

Sl. no	Type of error	Definition
1.	Missing code error	$x(n) - x(n-1) = 2$ LSB
2.	Monotonicity error	$x(n) - x(n-1) =$ negative
3.	Differential non-linearity error	$x(n-1) = x(n) = x(n+1)$
4.	Integral non-linearity error	Ideal(n) $- x(n) >$ boundary

$W_{Current}$ - $W_{pevious}$ = 1LSB -- Missing code error
$W_{Current}$ - $W_{pevious}$ = negative (for increasing input) ----------- Monotonocity error

Figure 2.10. Different types of errors.

decreases even when the input signal is monotonically increased. This type of fault is defined as a monotonicity error. If the ADC generates a sample of the same value for more than three consecutive samples, $x(n-1) = x(n) = x(n+1)$, this means that the ADC output remains constant even when the input signal is increased by 3 LSBs. This fault corresponds to the DNL error. Ideally, if the outputs of two consecutive samples are identical, then the ADC under test can be considered as a faulty circuit. However, due to some uncertainty in analog circuits, the boundary of the DNL error should be relaxed to at least three samples. The INL/gain error is defined by the difference between the ADC's output and the ideal output that is obtained by counting the clocks in the ramp signal generator. The INL/gain error can be caused either by the non-linearity or by the gain error of the ADC under test.

The missing code error and monotonicity error are defined in figure 2.10.

2.7.2 Implementation of a code-width BIST scheme for testing pipelined ADCs

A code-width BIST architecture is shown in figure 2.11. Initially, we began implementing this using MATLAB Simulink. Here a code-width type test is implemented on an 8 bit pipelined ADC. To implement this a register is designed using Simulink to store the previous digital output, which is required to find out the difference between the two consecutive codes. This difference helps us to find the missing code error and monotonicity error. If the difference between the consecutive codes is more than 1 LSB then a missing code error results, which can be detected with a digital subtractor. If the code difference is negative for increasing input signal (ramp) then it results in a monotonicity error [11].

The faults can be tested using the schematic Simulink of MATLAB in figure 2.12.

The result of the subtractor output from the above schematic is displayed in figure 2.13.

From the above result it can be concluded that the code difference between consecutive codes is exactly equal to 1 LSB, so the missing code error can be detected if this difference is greater than 1 LSB. This is the test for the missing code error.

It can also be observed from figure 2.13 that the code difference is positive. Hence there is no monotonicity error. If the difference is negative then it results in a monotonicity error, which can be tested from this.

The faults, including DNL and INL error, can be detected by designing a counter which counts the number of clock pulses present in each code. From this these two tests can be detected.

2.7.3 An 8 bit edge triggered register

An 8 bit edge triggered register is designed in Cadence, which is a highly essential block in storing the digital bits of the previous code, from which the difference

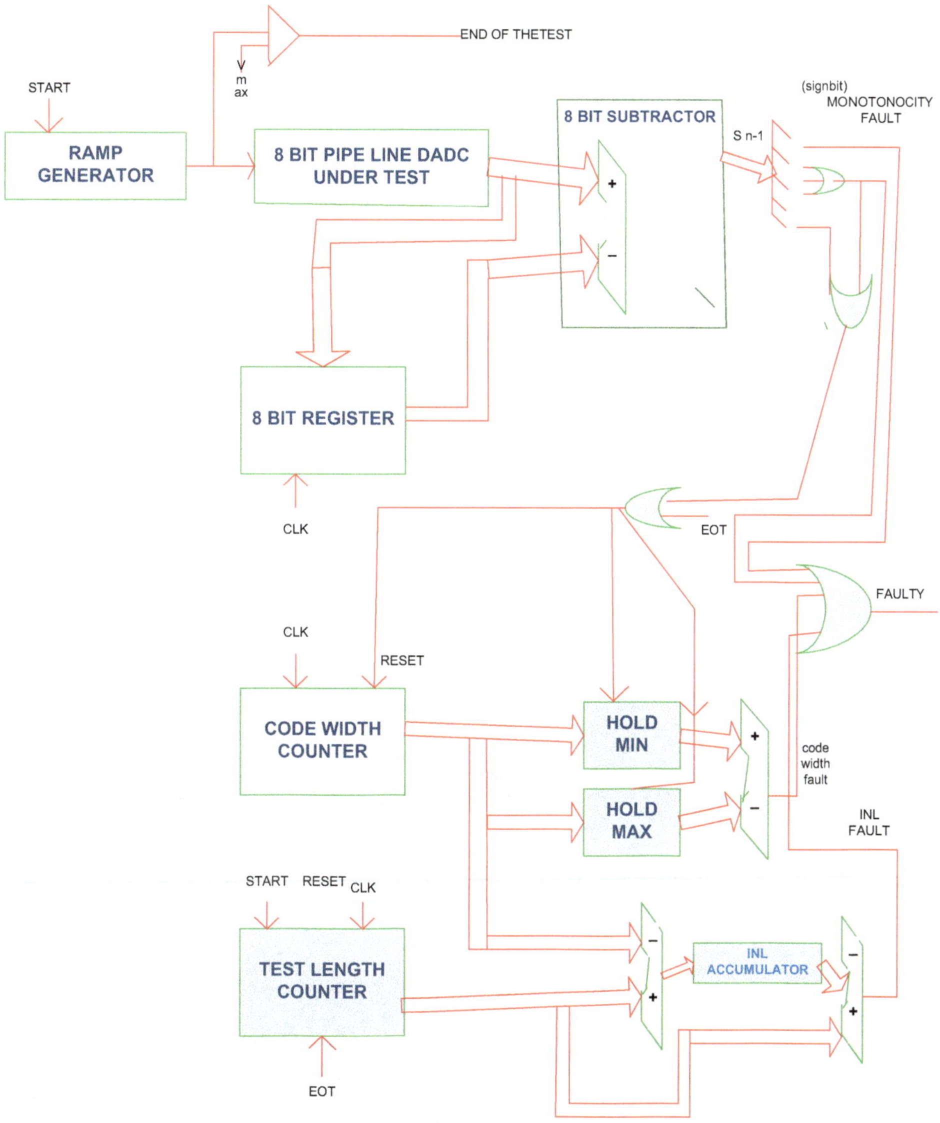

Figure 2.11. The code-width BIST architecture.

between consecutive codes can be calculated. This is designed using transmission gates as the switches and inverters for buffering. The schematic of the 8 bit edge triggered register is shown in figure 2.14.

The results of 8 bit digital output and corresponding analog output obtained by passing through the DAC from the register are shown in figure 2.15.

The first waveform corresponds to the analog output obtained by passing the digital output through the DAC. The other waveforms indicate the 8 bit digital outputs from the register. The digital output which is obtained from the register

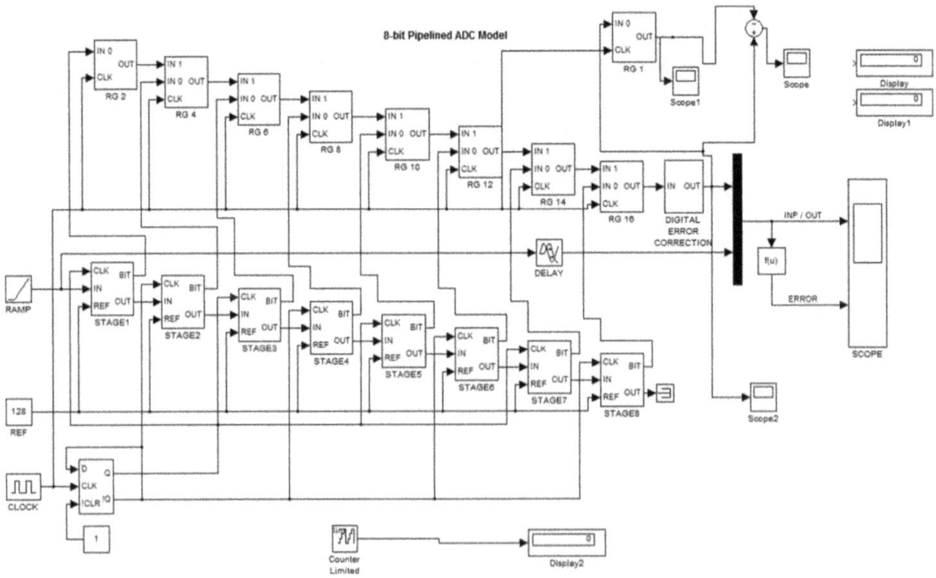

Figure 2.12. Schematic of the code-width testing scheme simulated using Simulink.

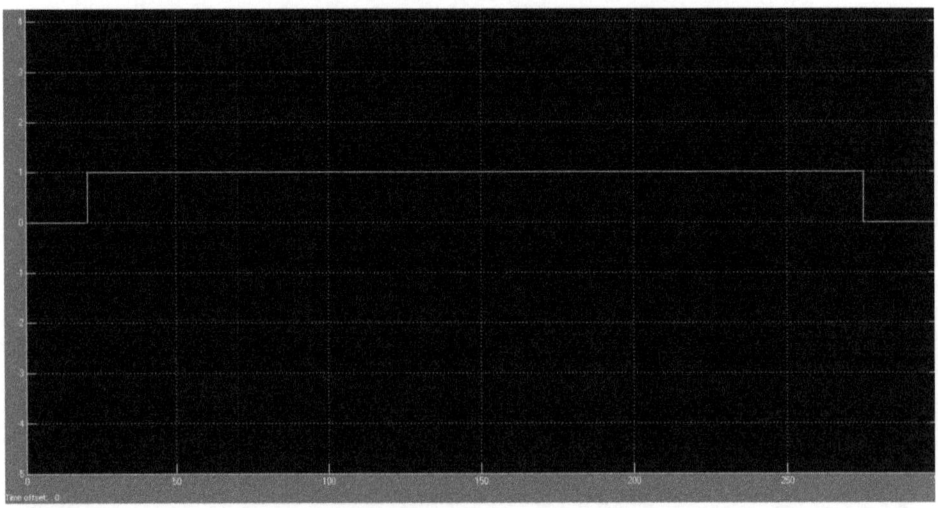

Figure 2.13. The result of the subtractor output using Simulink.

corresponds to the previous code which lags behind the current digital code by 1 LSB (one clock pulse).

2.7.4 An 8 bit subtractor

The design of an 8 bit subtractor is performed in Cadence, which gives us the result corresponding to the difference between the current digital code and the previous digital code. The schematic of the 8 bit subtractor is shown in figure 2.16.

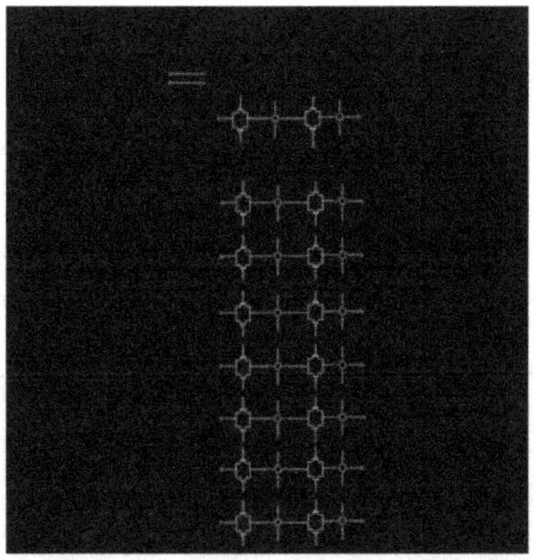

Figure 2.14. Schematic of the 8 bit edge triggered register.

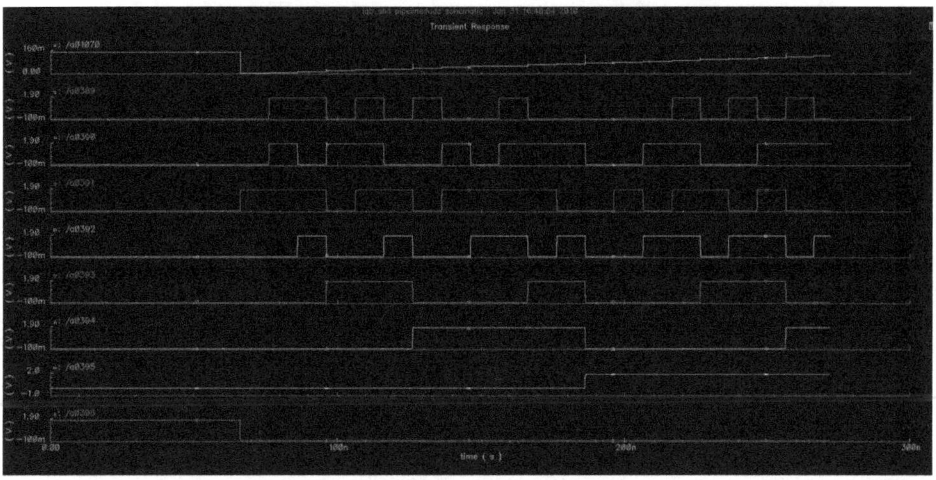

Figure 2.15. The results of 8 bit digital output and corresponding analog output obtained by passing through the DAC from the register.

It has 16 inputs, among which first eight inputs are given from the digital bits corresponding to the present code of the pipelined ADC and the next 8 bits are given from the register outputs, as shown in figure 2.17.

The results showing the analog output obtained from the pipelined ADC, the 8 bit register and the output from the subtractor unit are shown in figure 2.18.

In figure 2.18 the blue waveform represents the analog output obtained from the pipelined ADC when the eight digital bits of the pipelined ADC are passed through an 8 bit DAC. This corresponds to the current code of the pipelined ADC. The pink waveform represents the output obtained from the register when passed through the DAC. It can be observed that this lags behind the blue waveform by 10 ns, which corresponds to the code width of one LSB. The green waveform corresponds to the output obtained from the 8 bit subtractor and when passed through the DAC (the analog output).

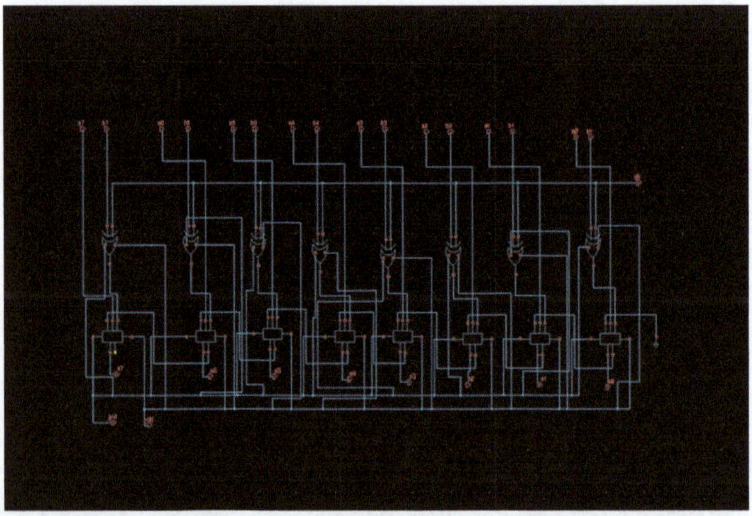

Figure 2.16. Schematic of the 8 bit subtractor.

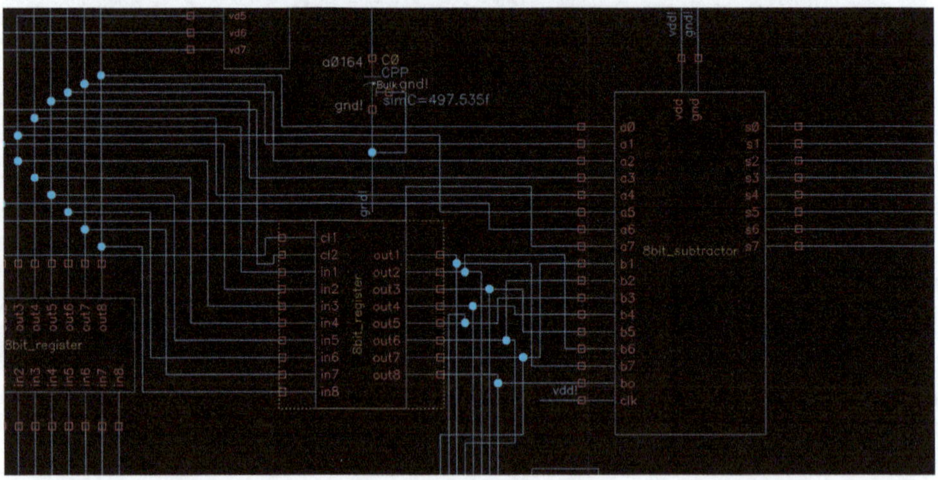

Figure 2.17. Circuit schematic of proposed pipelined ADC showing the number of bits distribution.

Note: It is clearly observed from figure 2.18 that the output obtained from the subtractor is exactly 1 LSB, which is shown. If it is more than 1 LSB then it results in the missing code fault, which can be detected using a comparator for comparing the subtractor output with the value corresponding to one LSB. If the subtractor output is negative then it results in the missing code fault, which can be detected by the comparator.

The spikes obtained in figure 2.19 are due to clock pulses that can be reduced.

2.8 Simulation of the pipelined ADC

Simulation of the pipelined ADC is shown in figure 2.20.

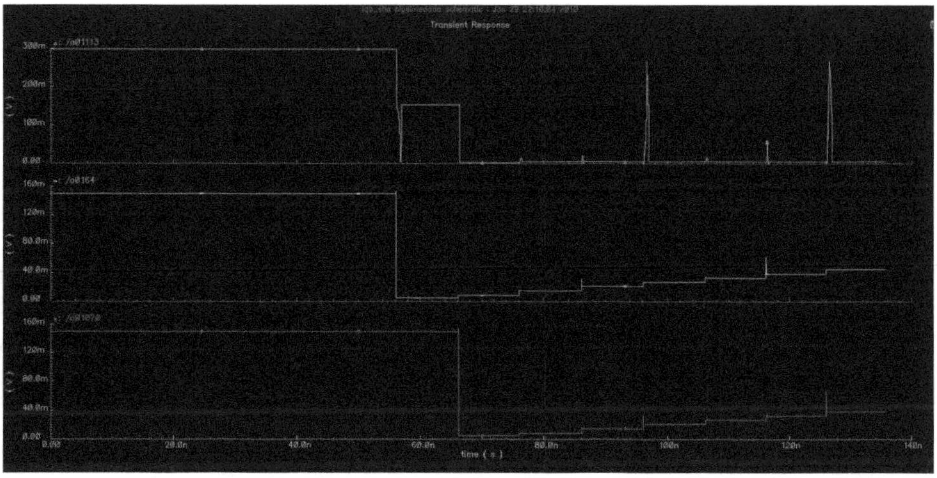

Figure 2.18. Analog output obtained from the pipelined ADC with 8 bit register and the output from the subtractor.

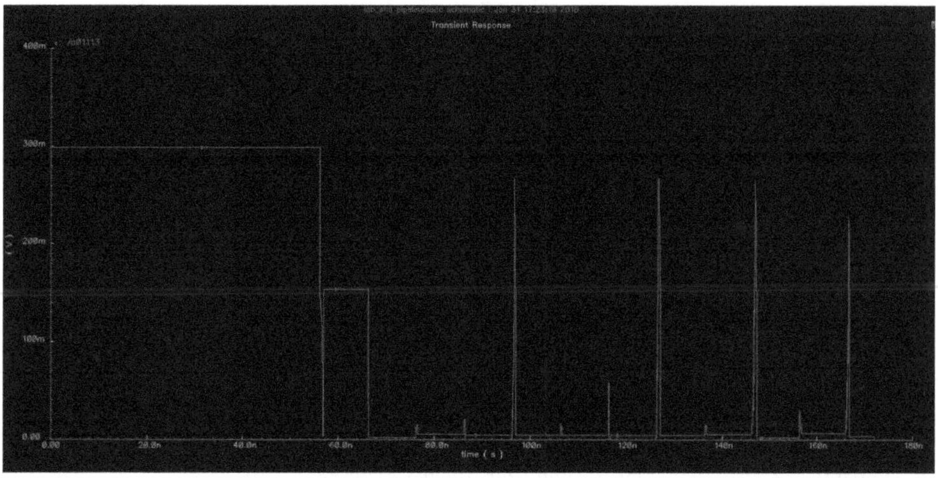

Figure 2.19. Subtractor output.

2.9 Future work

- Designing an 8 bit synchronous counter which is required for measuring the number of clock pulses generated in each code. This will help us to obtain the code width corresponding to each code. Using this we can test for the faults of DNL error and INL error.
- Designing an accumulator for storing the count which is required for detecting the INL error fault.

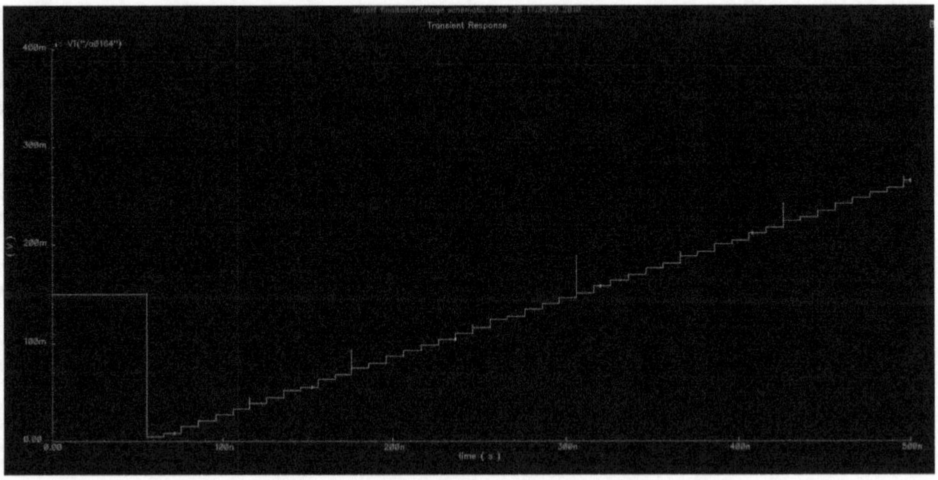

Figure 2.20. Simulation of the pipelined ADC.

References

[1] Behzad R 2002 *Design of Analog CMOS Integrated Circuits* (New York: McGraw-Hill)

[2] Allen P E and Holberg D R 2002 *CMOS Analog Circuit Design* (Oxford: Oxford University Press)

[3] Gray P R, Hurst P J, Lewis S H and Meyer R G 2008 *Analysis and Design of Analog Integrated Circuits* 4th edn (India: Wiley)

[4] Li J and moon U-K 2003 Background calibration techniques for multistage pipelined ADCs with digital redundancy *IEEE Trans. Circuits Syst.—II: Analog Digit. Signal Process.* **50** 531–8

[5] E-Waltari M and Halonen K A I 2002 *Circuit Techniques for Low-voltage and High Speed A/D Converters* (Dordrecht: Kluwer)

[6] Moon U K and Song B-S 1997 Background digital calibrated 12-bit, 10 M/s, 3.3 V A/D converter *IEEE Trans. Circuits Syst.—II: Analog Digit. Signal Process.* **44** 102–9

[7] Razavi B 1997 Design of sample and hold amplifiers for high-speed low voltage A/D converters *Proc. IEEE 1997 Custom Integrated Circuit Conf.* pp 59–66

[8] Shingawa M and Akazawa Y 1990 Jitter analysis of high-speed sampling systems *IEEE J. Solid-State Circuits* **25** 220–4

[9] Kabisatpathy P, Barua A and Sinha S 2005 *Fault Diagnosis of Analog Integrated Circuits* (Dordrecht: Springer)

[10] Wang J, Sanchez Sinencio E and Maloberti F 2000 Very linear ramp generators for high resolution ADC BIST and calibration *Proc. of 43rd IEEE Midwest Symp. on Circuits and Systems (Lansing, MI, August 8–11, 2000)* pp 908–11

[11] Barua A and Tausiff M 2011 A code width built-in-self-test circuit for 8-bit pipelined ADC *Proc. of 21st Int. Conf. on System Engineering (Las Vegas, NV, 16–18 August 2011)* pp 287–91

Chapter 3

Design of an oscillation-based built-in self-test system for a 1.8 V, 8 bit, 125 mega samples per second pipelined analog-to-digital converter

Aniruddha Biswas and Alok Barua

An oscillation-based built-in self-test (BIST) scheme is proposed for static testing of a pipelined analog-to-digital converter (ADC). In the BIST structure, the input signal to the pipelined ADC is controlled to oscillate linearly between two transition voltages and subsequently the ADC output oscillates between two digital codes, centering on a particular digital output. The period of oscillation is proportional to the code width of the output code and the ADC oscillates around that frequency (inverse of period). By counting the number of clock cycles during the rise or fall time of the oscillation and comparing this with the ideal number of clock periods corresponding to the ideal code width, the offset error, differential non-linearity (DNL), integral non-linearity (INL) and gain error can be obtained. The 'control logic' block is implemented to take care of the inherent delay of pipelined ADCs. Oscillation strategies are modified to make them fit pipelined ADCs. A delay sub-block is developed which disables the oscillation for an initial period equal to the inherent delay of the ADC, preventing erroneous results. An on-chip reference code generator circuit is made which sequentially changes the reference code, for which the ADC nonlinearities are measured. A missing code error detection block has been developed which is capable of measuring the missing code error during ADC oscillation. Monotonicity error is measured by disabling the oscillation and increasing the input to the ADC monotonically. For dynamic testing of the ADC, an oversampling-based analog sine generator has been designed.

doi:10.1088/978-0-7503-1732-0ch3

3.1 Introduction

3.1.1 Introduction to the BIST

Testing is a very crucial phase in a very large-scale integration (VLSI) chip's lifecycle. During the design phase testing is performed to detect and identify design errors, which is called design for test (DfT), while in the manufacturing phase testing is performed to detect manufacturing defects, which is called fault diagnosis. Finally, during system operation, testing seeks to detect any fault sustained during operation that would produce erroneous operation of that system. To ensure that the best product reaches the customer, rigorous testing methods are adopted during the product lifecycle. However, the cost of testing and the test time are the two major factors that come into play when stringent test requirements are to be satisfied and they definitely add to the product cost. This issue is very important in achieving high yield in VLSI circuit production.

Most system-on-chip (SOC) devices manufactured today contain some analog functionality leading to mixed signal chips with a large digital part and a small analog part. The integration of highly complex mixed signal systems on a single chip is becoming an industry standard.

With the advancement of complex mixed signal integrated circuits (ICs), faster and more complex test equipment is required to meet the stringent test specifications. Automatic test equipment (ATE) systems with high speed, precision, memory and performance are becoming increasingly expensive. For such a component density in SOCs, the test cost can be as high as 55% of the total manufacturing cost [1]. In addition, in ATE-based test schemes the circuits must be tested sequentially so the test time increases, which also has a large impact on chip cost. Reducing the ATE test costs and test times, applying simpler and faster ATE-based tests, and using less expensive test equipment are necessary to reduce the costs of testing. In addition, in highly complex mixed signal circuits, there are severe restrictions on the accessibility to the various nodes of the circuit under test (CUT). Moreover, there is also a limitation on the number of pins. The creation of extra test pins makes the chip bulky and more expensive. Hence it is becoming increasingly impossible to provide the test stimulus and perform the measurements outside the chip.

The above factors are pushing manufacturers to opt for self-test strategies where the requirements on the test equipment are reduced and the test circuitry is placed beside the CUT itself. This drastically reduces the test costs and also alleviates the problem of accessibility or the creation of more test pins at the cost of additional silicon area. This test scheme is called the built-in self-test (BIST) system and is a challenging current research topic. The basic essence of the BIST scheme is to design a circuit so that it can test itself and determine whether the CUT is faulty or not. Hence additional circuitry and functionality need to be incorporated into the design to facilitate the self-testing feature. This additional functionality must be capable of generating the test stimulus as well as analyzing the response of the CUT to the test stimulus to determine whether the CUT is faulty or fault-free.

Analog-to-digital converters (ADCs) are a very important component of mixed signal circuits. Efforts are being made to develop new ADC designs which will ensure high sampling speed, high accuracy and low power. Hence new test algorithms should also be developed to test high performance ADCs. Therefore, new BIST systems are being developed for these high performance ADCs with different architectures.

The testing of ADCs can be classified into static testing and dynamic testing [1]. In static testing a slowly varying (low frequency) signal is applied and the following test parameters are evaluated from its output response:

1. Offset error.
2. Differential non-linearity (DNL).
3. Integral non-linearity (INL).
4. Gain error.
5. Missing code error.
6. Monotonicity error.

For dynamic testing of ADCs a high frequency sinusoidal signal is applied to the ADC and from the output response the following parameters are evaluated:

1. Signal-to-noise ratio (SNR).
2. Total harmonic distortion (THD).
3. Spurious free dynamic range (SFDR).
4. Effective number of bits (ENOB).

An oscillation-based BIST system has been designed for the static testing of a 1.8 V, 8 bit, 125 MSPS pipelined ADC. In this test scheme the ADC under test is made to oscillate around an output code and from the frequency of oscillation the static performance parameters, mentioned above, are evaluated. In this chapter an oscillation-based BIST is reported. The test stimulus generator and response analyzer circuits are created. A 'control logic' block is produced which controls the total testing process. It creates the ADC oscillation and also enables the error measurement blocks.

The characteristic of a pipelined ADC is its inherent latency. When an analog input is applied to the ADC, the corresponding digital code appears at the ADC output after a delay, after being processed sequentially by the number of pipelined stages. Due to the inherent delay of the pipelined ADCs, test blocks may give erroneous results. So the 'control logic' block has been modified to start the ADC oscillation after the ADC inherent delay period and also to disable the error measurement block for that initial period.

Again, for the dynamic testing of the ADC, an oversampling-based sinusoidal signal generator has been designed to produce a spectrally pure sine wave. Instead of sticking to conventional analog sine generation principles, a lossless discrete integrator (LDI)-based digital sine generator has been developed, which is followed by a digital-to-analog converter (DAC) and a low-pass filter. A delta-sigma loop has been used in the oscillator loop to reduce the huge area required by a basic LDI-based oscillator.

3.1.2 Literature review

Several BIST techniques are reported for static and dynamic testing of ADCs. A very popular method for testing ADC static errors is the histogram test technique [2–4]. It involves the application of an analog signal to the ADC input and recording the number of times each code appears at the ADC output. The analog signal can be any function whose amplitude distribution is known. ADC errors modify the output code count and hence the histogram shape. Hence measuring the output code count and comparing it to the ideal count and performing some complex calculations permits one to evaluate the offset error, gain error, DNL and INL. The triangular or sinusoidal signal is given as the input to the ADC. However, this BIST scheme requires a large amount of hardware resources. Hardware minimization of histogram BIST for a sinusoidal input signal has been performed in [5, 6]. For the histogram BIST scheme the test time is a major issue for high resolution ADCs (test time $= \frac{72}{F_S} 2^{2n}$, where F_s = sampling frequency and n = ADC resolution) [6].

A BIST structure is proposed in [7] which permits one to evaluate the converter linearity by using only the least significant bit (LSB) of the digital output, the global functionality being tested with the comparison between the remaining bits and a counter clocked by the LSB. In the code-width-based static testing scheme, the slope calibrated input signal is not required. The test circuit is a simple digital circuit with a gate count of only 550 and the simulated fault coverage is approximately 99% [8].

A fully digital test approach is proposed in [9]. Here the main advantage of the BIST circuit is that it can test DNL and INL for all codes in the digital domain and does not need any calibration. Again, here some parts of the ADC under test are used in the BIST with minor modifications, which reduces the chip area. The proposed BIST structure presents a compromise between test accuracy, area overhead and test costs. Another BIST scheme is proposed in [10], where delta-sigma-based linear ramp generation is used as a test stimulus and is applied to the ADC under test and subsequently the DNL and INL of the converter are measured. Here 5% LSB test accuracy can be achieved in the presence of reasonable analog imperfection. The main advantage of this scheme is that it does not require both on-chip ADCs and DACs, which makes it suitable for most mixed signal ICs. Again, both the stimulus generation and measurement techniques are highly tolerant to analog variations. Another proposed BIST scheme is based on sequential code analysis [11]. In addition to the measurement of DNL and INL, non-monotonic behavior of ADC can also be detected. A ramp generator is used to generate test stimuli. The feedback configuration is used in the implementation of a ramp generator to increase linearity.

A low cost BIST solution for static and dynamic testing of ADC is provided in [12]. Here the pseudo-noise signal has been given as the test stimulus. The silicon area requirement and test time are also reduced in comparison to the histogram method. In a separate method the dynamic performance of ADCs (THD and SFDR) is estimated from the INL data achieved from static characterization of the ADC, without requiring additional data acquisition or accurate sinusoidal sources

[13]. Some earlier works on oscillation-based BIST have been reported in [14] and [15].

Analog sinusoidal wave generation is shown using direct digital frequency synthesizers (DDFS) in [16] and [17]. But this approach is chip area intensive. An oversampling-based sinusoidal oscillator is presented in [18] and [19]. With the exception of a low-pass filter and a 1 bit DAC, the circuit is entirely digital ensuring accurate control over oscillation frequency and amplitude. To decrease circuit complexity the entire oscillator is operated at oversampling frequency. The incorporation of the delta-sigma loop inside the oscillator loop ensures a very efficient implementation of the oscillator requiring a minimum silicon area. Finally, a time domain approach for testing the SNDR and ENOB of an ADC is illustrated in [20], which does not rely on the conventional fast Fourier transform (FFT) method approach to determine these parameters from the frequency spectrum.

3.1.3 Motivation and aims

The oscillation-based BIST method is a general test scheme applied to functional and structural testing of mixed signal circuits. The basic principle of this test scheme is that the circuit under test is reconfigured to oscillate and important information regarding the functional and structural parameters of the circuit is extracted from the oscillation frequency and is stored as a healthy signature. The deviation of the frequency of oscillation from its nominal value indicates a faulty circuit. The oscillation-based test strategy has been successfully applied to a wide range of analog and mixed signal circuits, including analog filters and operational amplifiers. However, the application of this test strategy has never been tested for functional testing of pipelined ADCs.

With the motivation above, the goal of this work is to apply the oscillation-based BIST (OBIST) methodology for functional testing of pipelined ADCs. A 1.8 V, 8 bit, 125 MSPS pipelined ADC is used as the CUT to test the OBIST methodology and subsequently to evaluate the static performance parameters listed below:

1. Offset error.
2. DNL.
3. INL.
4. Gain error.
5. Monotonicity error.
6. Missing code error.

In addition to further testing of dynamic performance parameters, a delta-sigma modulator-based oversampling sinusoidal signal generator is built, ensuring minimum silicon area.

3.1.4 Brief description of the work

An oscillation-based BIST system is developed for a 1.8 V, 8 bit, 125 MSPS pipelined ADC, which was designed by previous workers [21, 22]. The on-chip test circuit is designed and simulated in the Cadence platform with standard 180 nm

CMOS technology. The oscillation-based BIST scheme extracts the code width of every possible ADC output code and compares it with the ideal code width to obtain the linearity errors. The ADC is fed from an on-chip ramp generator. The ramp generator is controlled to oscillate between two voltage transition edges by a 'control logic' block corresponding to two ADC output codes. The 'control logic' block is designed, which initiates and maintains the oscillation and is the heart of the test scheme. A control signal from the 'control logic' block makes the signal generator ramp up and ramp down sequentially and hence the ADC oscillates around a reference code. The code width is proportional to the increasing or decreasing time of the oscillating signal. The increasing or decreasing time is obtained by counting the number of clock cycles in that period. It is compared with the ideal number of clock counts (according to ideal code width) to obtain the offset error, DNL, INL and gain error. Separate measurement blocks are developed to test the offset error, DNL, INL and gain error. Missing code error and monotonicity error detection blocks are designed and included in the oscillation-based test strategy.

The major structural blocks of the OBIST system are modified from previous works on ADC testing, to incorporate the inherent delay of the pipelined ADCs. A separate 'delay sub-block' is incorporated in the 'control logic' block which initiates the oscillation after the initial delay period of the ADC is over. Also, for this initial period all the measurement blocks are disabled by the 'control logic' block. Oscillation strategies are modified inside the 'control logic' block to ensure error-free testing of pipelined ADCs.

For the dynamic performance testing of the ADC, a delta-sigma modulator-based sinusoidal signal generator is developed. The same is simulated in MATLAB. The output sine wave is spectrally pure with an SNR of 123 dB. The oversampling-based sinusoidal signal generator is then designed in the Cadence environment with standard 180 nm technology.

3.1.5 Chapter organization

This chapter is organized as follows. In section 3.2 the concepts behind a general BIST scheme are discussed. The general principle for an oscillation-based BIST system is presented, which is applicable to analog and mixed signal circuits. Then the oscillation-based BIST principle and BIST structure for functional testing of ADCs are discussed in detail. The determination of static performance parameters from the BIST scheme is presented along with a detailed BIST structure. The main challenges in BIST design for pipelined ADC are also discussed here.

In section 3.3 the circuit implementation of all the major blocks of the OBIST system are presented along with the intricacies and design constraints. The simulation results for each block are also presented and discussed. In section 3.4, the dynamic BIST for ADCs is introduced. The concepts behind the design of an area-efficient analog oscillator are discussed. The simulation results for the oscillator are also presented. In section 3.5, conclusions and a brief summary of the work are presented. Finally, the future scope of the work is discussed.

3.2 Oscillation-based BIST principles

3.2.1 General BIST principles

The basic idea of BIST is to design a circuit so that it can test itself and determine whether the CUT is faulty or healthy. This typically requires that additional circuitry and functionality are incorporated into the design of the circuit to add self-testing features. This additional functionality must be capable of generating test patterns or test stimulus as well as providing a mechanism to determine if the output response of the CUT to the test patterns corresponds to that of a fault-free circuit. So BIST is basically a method of testing an IC that uses special circuits built into the IC. This circuitry performs test functions on the IC and signals whether the parts of the IC covered by the BIST circuit are working properly.

3.2.2 BIST basic test flow

The basic test flow of BIST is shown in figure 3.1. A determined set of input stimuli is applied to the CUT and the output response of the circuit corresponding to the test stimuli is compared to a known good response or expected response to determine if the circuit is good or faulty. There may be feature extractor circuits which extract important information from the output response and then compare them to expected test parameters.

3.2.3 Basic BIST architecture

A general BIST architecture is illustrated in figure 3.2.

In figure 3.2 the test stimulus generator produces the input stimuli to the CUT and the output response analyzer extracts important information and test parameters from the response of the CUT to the test stimuli and generates a 'healthy' or 'faulty circuit' signal after comparison with the nominal or normal test parameters

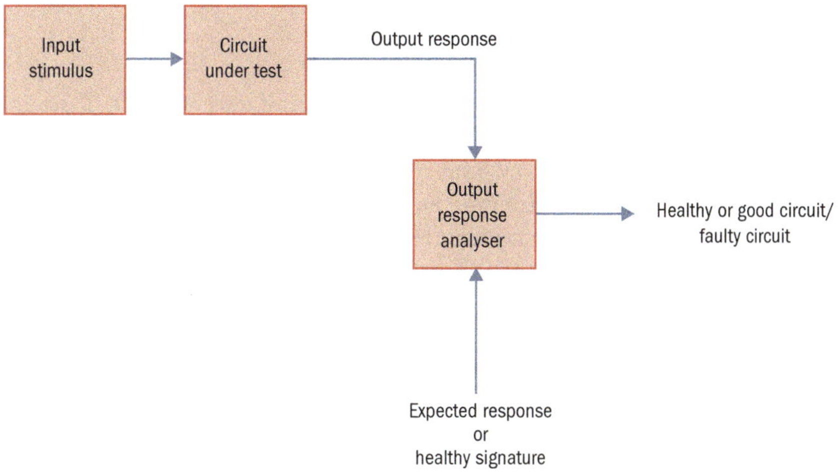

Figure 3.1. BIST basic test flow.

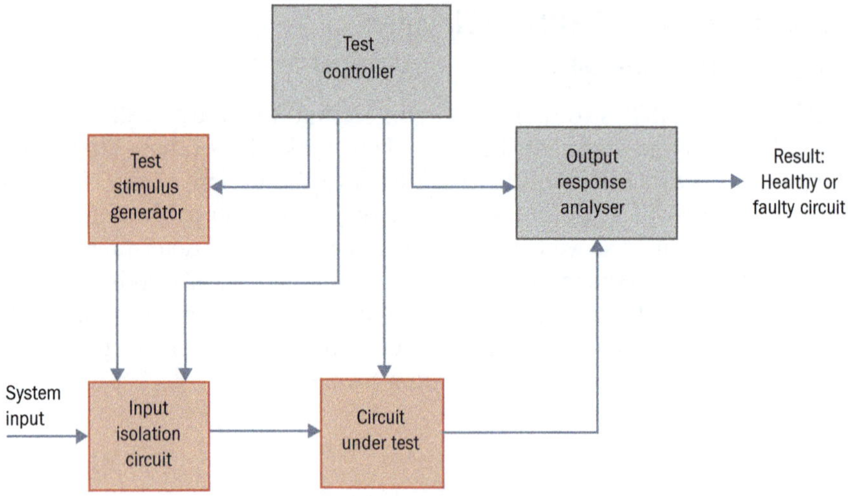

Figure 3.2. General BIST architecture.

corresponding to a fault-free circuit. The total operation is controlled by the 'test controller' block which is the crux of the test system. The entire time sequence control is performed by the test controller. The initialization of the CUT and the BIST system is also performed by the test controller before starting the test to ensure reproducibility. The input isolation unit ensures that the system input should not be applied to the CUT while the testing is in process. Only the test stimulus generator input should be given to the CUT. This is generally achieved with multiplexers or blocking gates. Apart from normal system input/output pins, incorporation of BIST may also require additional input/output pins for activating the BIST sequence (BIST start), reporting the results of the BIST (pass/fail indication) and an optional indication that the BIST sequence is completed, the BIST results are valid and can be read to check the status of the CUT (healthy or faulty).

3.2.4 Principles of oscillation-based built-in self-test

The basic idea of the OBIST approach is to reconfigure the CUT to form an oscillating circuit and to evaluate the frequency of oscillation. The frequency of oscillation is related to functional and/or structural parameters of the CUT. The oscillation frequency can be considered as a digital signal and can be easily evaluated using digital circuitry, which increases the precision of the test.

In figure 3.3 [14] the general OBIST structure is shown which is applicable for the functional and structural testing of analog and mixed signal circuits. The major blocks in the generalized structure are the analog multiplexer (AMUX), the frequency-to-number converter (FNC), the control logic block (CL) and the CUT. In the test mode the CUT is partitioned into appropriate sub-circuits, which are reconfigured as oscillators one after another. The AMUX selects the test point extracted from the converted building block. The oscillation frequency of the selected test point is then converted to an N-bit number by the FNC and is

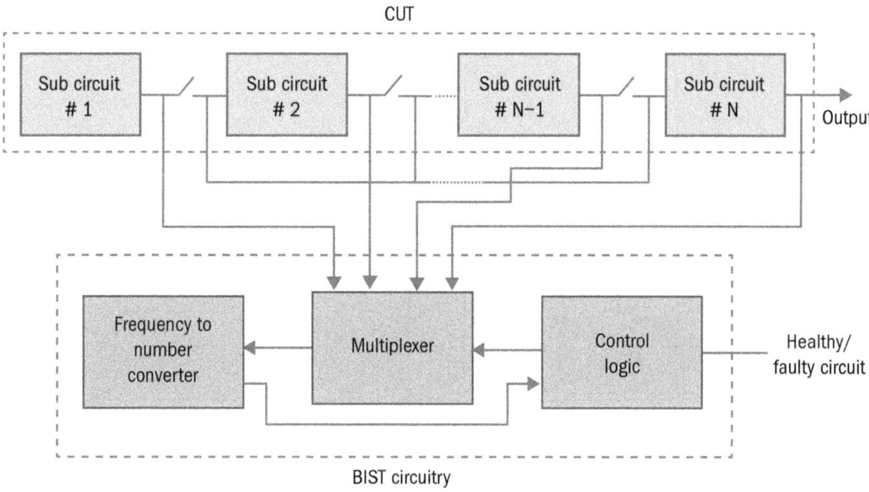

Figure 3.3. General OBIST structure.

evaluated by the control logic to generate the pass/fail (healthy or faulty circuit) status. The control logic block directs all the operations during the testing mode.

3.2.5 Oscillation BIST principles for functional testing of ADCs

The oscillation-based BIST scheme for functional testing of ADCs is based on obtaining the code width of each output code and comparing it with the ideal code width so as to obtain the static performance parameters. In this scheme the ADC (the CUT) is made to oscillate between two predetermined codes. The frequency of oscillation gives important information about the ADC. The deviation of the oscillation frequency from its nominal value indicates a faulty circuit.

The transfer characteristic of a typical ADC is shown in figure 3.4. In this test scheme the ADC is fed from a ramp signal source. For evaluating the code width for a particular output code C_j, the ADC input is made to oscillate linearly between input voltages V_j and V_{j+1}, where V_j is the transition voltage between code C_{j-1} and C_j, and V_{j+1} is the transition voltage between codes C_j and C_{j+1}. So if the input voltage of the test ADC is varied continuously between V_j and V_{j+1}, the output code of the ADC will oscillate between codes C_{j-1} and C_{j+1}, centering the code C_j. When the output code is C_{j-1}, the input of the ADC is made to increase linearly. However, when the output code is C_{j+1}, the input to the ADC is made to decrease linearly and thus when the ADC input oscillates between V_j and V_{j+1}, and the output code oscillates between C_{j-1} and C_{j+1}. As the input to the ADC is linearly increasing and decreasing, its period is proportional to the code width for a particular code. The increasing or decreasing period is evaluated by a digital up-counter and then compared with the ideal value (according to the ideal code width for error-free ADC), then the ADC linearity errors can be obtained. While increasing the input to ADC from '0' volts, if the time is counted up to the first code transition of the ADC and compared with the ideal time count, the 'offset error' of the ADC

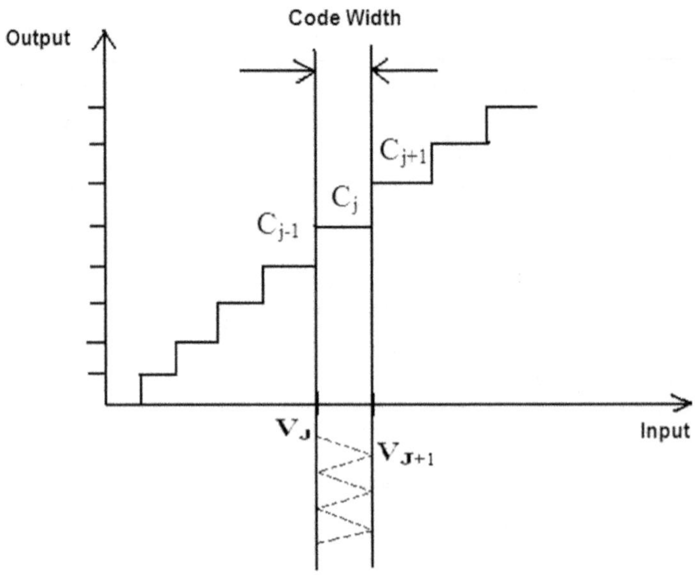

Figure 3.4. ADC transfer characteristics.

is obtained. While the input to the ADC is linearly increasing, if the two consecutive output codes are compared then the 'missing code error' can be obtained (when they are more than 1 LSBs apart and provided there is no monotonicity error). If the oscillation is disabled and the input to the ADC is made to increase monotonically and the two consecutive ADC output codes are compared, 'monotonicity error' can be obtained (if the output code decreases while the ADC input is increasing).

3.2.6 Oscillation-based BIST architecture

The basic block diagram of the oscillation-based BIST for static testing of ADCs is shown in figure 3.5. As can be seen, it is a loop structure with a feedback which creates the oscillation at the ADC output. As shown in the figure, the ADC is fed from a ramp generator. The ramp is generated by charging or discharging a capacitor through a constant current source. There is a switch in the generator circuit which determines whether the capacitor will be charged or discharged, or in other words the generator will ramp up or ramp down. This switch is controlled by a control signal from the control logic block. Hence the control logic block determines whether the ramp generator will ramp up or ramp down.

The output digital word of the ADC is fed to the control logic block. The input to the ADC is controlled by the control logic block through a control signal. This control signal makes the ramp generator ramp up or ramp down. This control signal also completes the loop by the negative feedback to create the oscillation. For evaluating the code width of a particular code C_j, the generator is made to ramp up

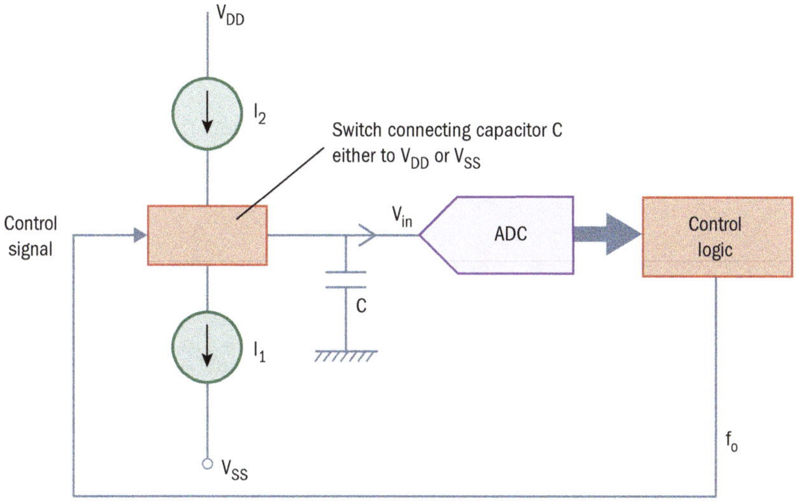

Figure 3.5. Oscillation-based BIST basic block diagram.

from '0' volts by the control logic block. The ADC output codes go on increasing as the input to the ADC ramps up. When the output of the ADC reaches C_{j+1}, the control signal from the control logic block makes the ramp generator ramp down. Again, when the output of the ADC reaches C_{j-1} (after being decreased from C_{j+1}), the control logic block directs the ramp generator to ramp up. In this way the oscillation of the ADC output codes continue centering code C_j. During the ramp-up period or the ramp-down period, if the number of clock cycles is counted, then it will be proportional to the code width of code C_j.

It can be observed that in this work for static testing of ADC, the use of an FNC is intentionally avoided, saving considerable silicon area. Despite measuring the frequency of oscillation and then converting it to an N-bit number using FNC, the code width is measured in terms of the number of clock counts during the occurrence of the reference code at the ADC output in the ramp-up or ramp-down period of the generator.

To verify the proportionality of the code width to the number of clock counts, the signal generator is to be discussed in more detail. It is well known that if a capacitor is charged with a constant current source, the voltage across the capacitor will rise linearly and generate a linear ramp. Similarly, if a charged capacitor is discharged by a constant source, the voltage across the capacitor will decrease linearly. This very simple principle is used to design the input stimulus circuit controlled by a control signal generated from the 'control logic'.

As shown in figure 3.6, the capacitor is charged and discharged by the ideal current sources I_2 and I_1, respectively. As shown in figure 3.6, the position of the CMOS switch will determine whether the capacitor will be charged or discharged. The position of the switch (in the charging position or in the discharging position) is governed by the control signal to oscillate the input. If the current sources I_2 and I_1 are equal, the slope of the charging and discharging will be equal. When the control

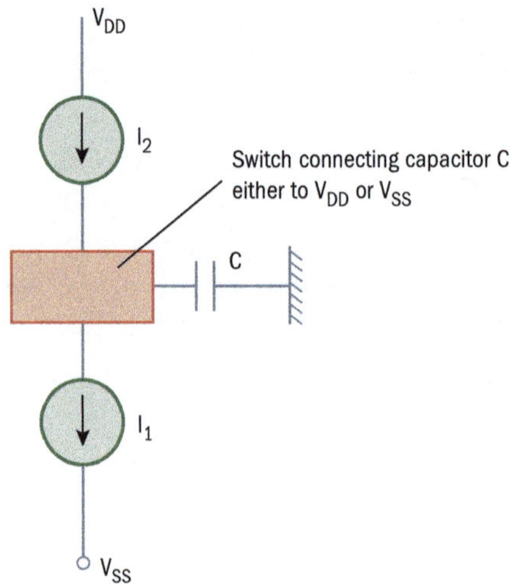

Figure 3.6. Signal generator.

signal is low the capacitor is charged and when the control signal is high the capacitor is discharged.

For an actual ADC let V_j and V_{j+1} be the transition edges of the code C_j, assuming that at time t_1 the voltage is V_j and at time t_2 the voltage is V_{j+1}, then the code width is

$$\Delta V = V_{j+1} - V_j = \frac{1}{C}\int_{t_1}^{t_2} Idt = \frac{I}{C}(t_2 - t_1) = \frac{I\Delta t}{C}. \tag{3.1}$$

Hence the code width is proportional to the clock count during the occurrence of the code at the ADC output.

3.2.7 Determination of the static performance parameters of an ADC using the OBIST scheme

In this subsection the testing methods of all DC or static parameters of ADC will be described.

Offset error: For an ideal ADC the output code is 0 if the input voltage is less than V_1, but for the actual ADC the transition voltage may be V_T. According to the definition of offset error

$$\text{offset error} = \frac{V_I - V_T}{CW_i}, \tag{3.2}$$

where CW_i is the ideal code width. The offset error test starts when the output code of the ADC is 0. A counter is started to count the number of reference clock pulses until the output of the ADC is greater than 0 and reaches its first transition. If the actual number of clock pulses is N_T and the ideal number of clock pulses is N_I, then

$$\text{offset error} = \frac{N_I - N_T}{N_I}. \tag{3.3}$$

DNL error: The input signal of the test ADC is oscillated between V_j and V_{j+1}, and the code is C_j. If we can count the number of clock pulses of a reference clock signal when the input is either increasing or decreasing, the number of clock pulses will be proportional to the actual code width C_j. If we compare it with the ideal number of clock pulses, we will obtain the DNL if normalized with respect to the ideal number of clock pulses related to 1 LSB. If the actual and ideal times are M_T and M_I, then

$$\text{DNL}(j) = \frac{M_I - M_T}{M_I}. \tag{3.4}$$

INL error: The INL error can be derived by adding the DNL errors. At a specific code C_j, the control logic measures the DNL error and then adds this to the INL error of the code C_{j-1} to obtain the INL error for the code C_j:

$$\text{INL}(j) = \sum\nolimits_{i=1}^{j} \text{DNL}(i). \tag{3.5}$$

Gain error: The INL error for the most significant code (1 1 1 1.....1) is the gain error:

$$\text{gain error} = \sum\nolimits_{i=1}^{2^n-2} \text{DNL}(i). \tag{3.6}$$

Missing code error: When the difference between the two consecutive output codes is greater than 1 LSB, there is a missing code error in the ADC. When the input to the ADC is increasing, the two consecutive output codes are compared to detect the missing code error.

Monotonicity error: When the input to the ADC is monotonically increased, if the output of the ADC decreases, then there is monotonicity error in the ADC. The oscillation-based BIST for ADC fails if there is monotonicity error in the ADC. So for testing monotonicity error, oscillation is disabled and the input signal to ADC is made to increase linearly.

3.2.8 Challenges in designing a BIST for a pipelined ADC

When the oscillation-based BIST system described in the previous section is applied to test the static performance parameters of the pipelined ADC, several problems are faced due to the inherent delay of the pipelined ADC. In a pipelined ADC, due to the existence of a number of pipelined stages, there is always a finite delay between the applied input and its corresponding digital output code.

Challenge 1: When the input of the ADC is increased from '0' V, initially for a few clock periods the ADC output codes are of no use due to the inherent delay of the ADC. The output corresponding to '0' V comes only after the inherent delay of a few clock cycles. Hence erroneous results are obvious in the measurement of offset error, linearity errors and missing code errors if they are measured in this period.

Modifications to resolve the issue: To take care of this problem, the oscillation should be disabled for this initial delay period and the input to the ADC is made to monotonically increase. For this purpose a delay sub-block is incorporated in the control logic block, which forces the control signal to remain in the 'low' state for the inherent delay period of the ADC. Once this period is over, the delay sub-block enables the oscillation and also the error measurement blocks.

Challenge 2: When the output code of the ADC is C_{j-1}, the control signal is made to be low and the input to the ADC is increased. But due to the delay in the ADC, the codes go on decreasing below C_{j-1}. Similarly when the 'control logic' detects C_{j+1}, it makes the input of the ADC decrease, but due to the delay in the ADC the output code goes on increasing. Hence erroneous results are obvious in the measurement of linearity errors.

Modifications to resolve the issue: To take care of this problem, instead of switching the input at codes C_{j-1} and C_{j+1}, the switching is performed at reference code C_j itself and C_{j-1}. When the output of the ADC is increasing and reference code C_j appears, input to the ADC is made to ramp it down. But due to the inherent delay of the ADC code, C_{j+1} appears and then again decreases from it. The input ramp generator design should also be modified such that the slope of the ramp should ensure that only the code C_{j+1} appears and the higher codes do not appear.

3.2.9 OBIST full scheme

The full scheme of oscillation-based BIST for static testing of the pipelined ADC is shown in figure 3.7. The error measurement and detection blocks are also shown along with the oscillation loop.

The output of the pipelined ADC is fed to the control logic block and also to offset the measurement block and to the block for detection of missing code error and monotonicity error. The DNL measurement block and the INL and gain error measurement block are fed by the DNL_INL_EN signal coming from the control logic block. For a particular reference code these blocks count the number of clock cycles only during the time this signal is high.

During the testing of the monotonicity fault, the oscillation is disabled. Hence the control logic block sends an 'oscillation disable' signal status to the monotonicity and missing code error detection block. Hence monotonicity error measurement is only enabled when the 'oscillation disable' signal is high.

3.3 Implementation of oscillation-based BIST

The major blocks of the oscillation-based BIST scheme are designed in the Cadence platform in standard 180 nm technology. The 8 bit pipelined ADC to be tested was an already designed one with a V_{DD} of 1.8 V and a sampling speed of 125 mega samples per second (MSPS). The design of the major blocks is given below.

3.3.1 Signal generator

A CMOS ramp generator circuit is shown in figure 3.8. As mentioned above, the voltage across the capacitor increases or decreases linearly when a capacitor is

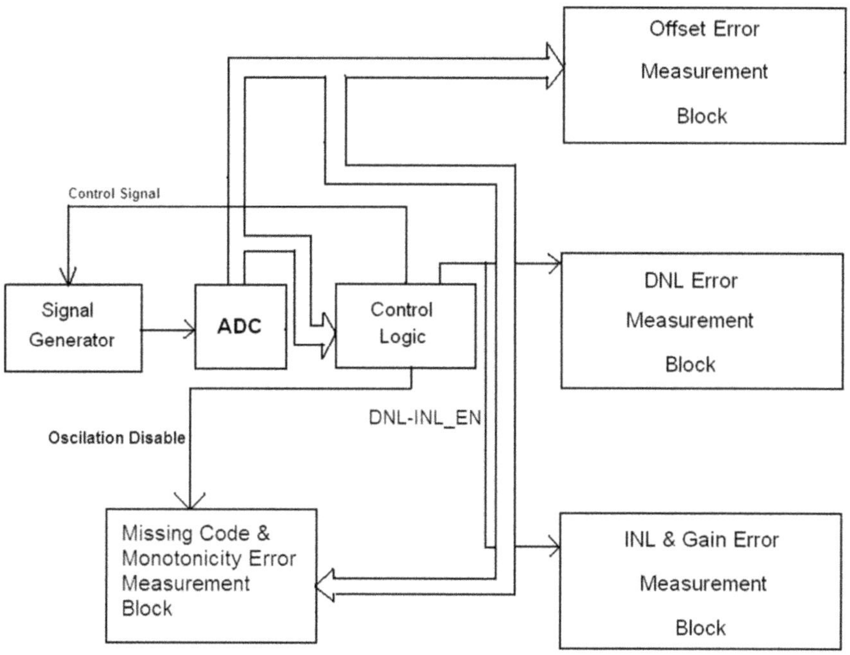

Figure 3.7. Oscillation-based BIST full scheme.

charged or discharged through a constant current source. Whether the capacitor will be charged or discharged will depend on the position of a switch, which is controlled by a control signal from the control logic block.

Since a MOSFET operating in the saturation region works like a constant current source independent of the voltage across the drain and source, an NMOS M_1 in saturation is used to charge the capacitor, as shown in figure 3.8. Now, to use the same constant value in the current source for discharging, current mirroring is used. Here M_1 and M_2 comprise a current mirror. M_7 carries the same current as M_2. Again M_7 and M_5 comprise the second current mirror. Hence the current flowing through M_1 is copied to M_5. The NMOS M_4 and PMOS M_3 together act as the switch which toggles the generator to the ramp-up mode and ramp-down mode. The control signal from the control logic block switches 'ON' M_3 when it is 'LOW' and hence the capacitor is charged linearly. Similarly, when the control signal is 'HIGH' it switches 'ON' M_4 and hence the capacitor is discharged linearly.

The design of the signal generator mainly involves the selection of the constant current source value and the capacitor value to control the slope of the ramp. The selection of the slope of the ramp is dependent on several factors which are as follows:

1. The slope should be slow enough such that several samples are taken for a change of 1 LSB at the output. If the slope is made higher, then there are chances that the output of the ADC misses some output code and delivers

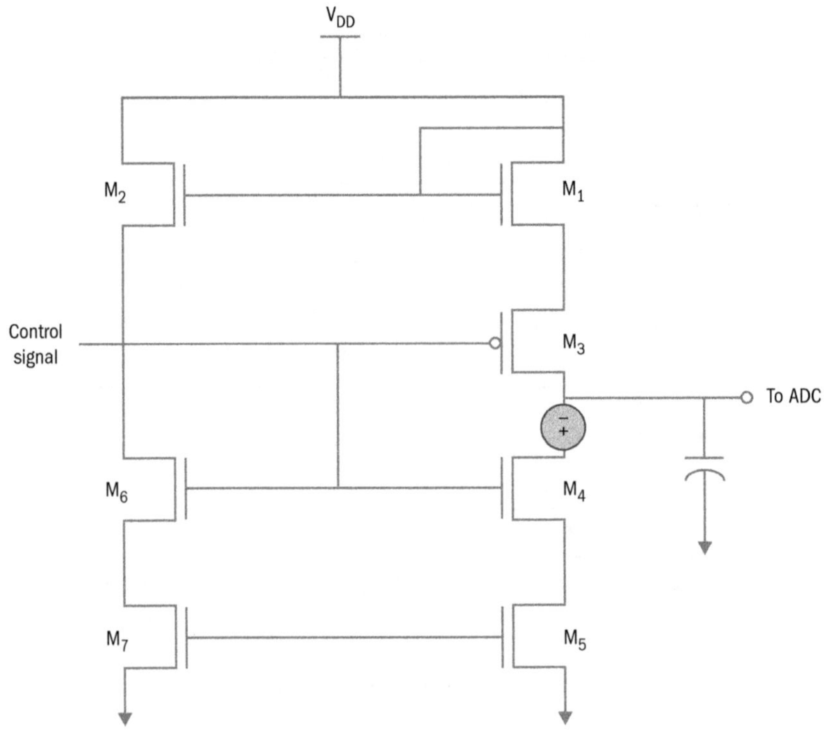

Figure 3.8. Ramp generator.

some output code which is greater than 1 LSB from the previous code. Hence the false missing code error alarm will be generated. Further, for good resolution for error calculation in the error measurement blocks (offset error, DNL, INL and gain error), the slope of the ramp should be low.

2. Again as we decrease the slope of the ramp, the test time increases.
3. For a reference code C_j, when the input to the pipelined ADC is increasing the control logic block makes the signal generator ramp down when the ADC output is C_j. Hence to make sure that the code C_{j+1} appears, the slope of the ADC should be increased. But the slope of the ramp should not be so high that a code higher than C_{j+1} appears at the ADC output. This will give erroneous test results. Similarly, the slope of the ADC should not be so low that while ramping down codes less than C_{j-1} appear, which increases the test time. Taking into account all of the above factors an optimum slope of the ramp has been selected. The ramp generator output is simulated by applying a square wave at the generator input (control signal). The simulated result is shown below in figure 3.9.

It is obvious from figure 3.9 that the slopes of the ramp-up and ramp-down modes are the same.

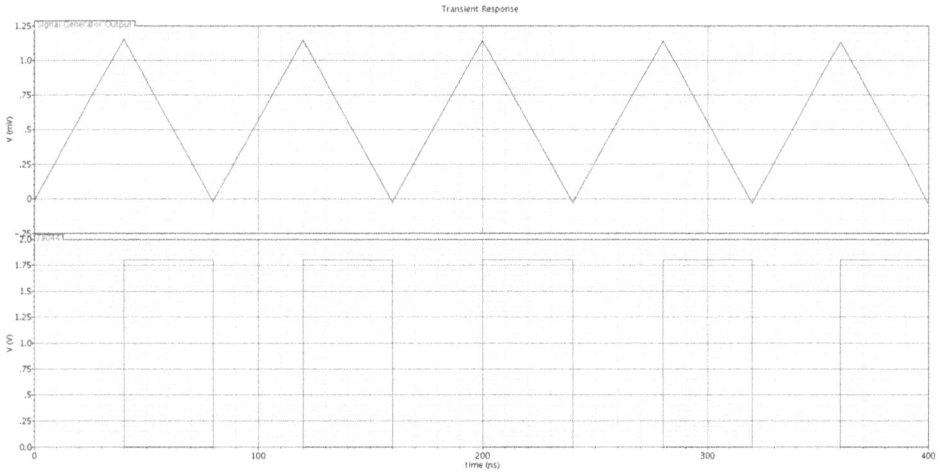

Figure 3.9. Signal generator output.

3.3.2 Control logic block

The control logic block controls the entire BIST system and hence it is the most important block of the BIST implementation. The control logic block is divided into four sub-blocks with different specific functionalities.

3.3.2.1 Sub-block for control signal generation

This part of the control logic block performs the very basic functionality, i.e. creation of the oscillation in the ADC and hence this is the crux of the control logic block. For a particular reference code C_j, this sub-block compares the ADC output code with the reference code (the output code around which the ADC oscillates) and makes the control signal high when the ADC output is equal to or more than the reference code. This is performed by a digital 8 bit subtractor unit using the two's complement method for subtraction, where the reference code is subtracted from the ADC output code. The two's complement of the reference code is taken and added to the ADC output code. When the ADC output code is lower than the reference code, no carry is generated. But when the ADC output is equal to or higher than the reference code, the carry is 'HIGH'. This carry is used as the control signal. The control logic block schematic is shown in figure 3.10.

3.3.2.2 Sub-block for reference code generation

For testing the DNL and INL errors for all the possible output codes of the pipelined ADC, it should be made to oscillate around each code and then the linearity errors are measured during the ramp-down period of oscillation. The reference generation sub-block generates the reference codes (around which the ADC has to oscillate) and linearity errors are to be measured automatically and sequentially from (0000…1) to (1111…1). To ensure the least test time, after the linearity measurement for a particular code, the signal generator is not reset and is

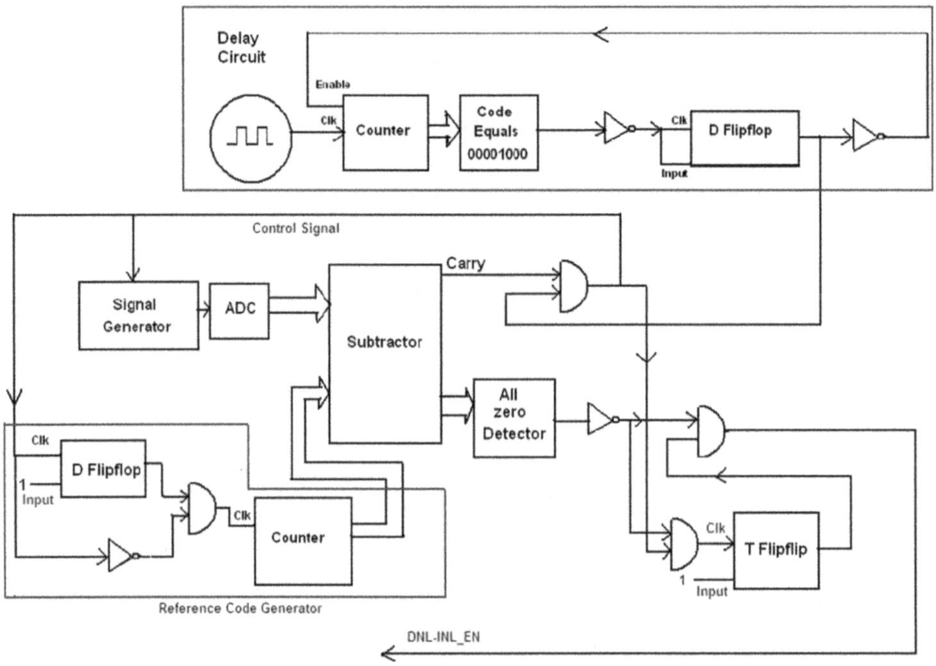

Figure 3.10. Control logic block.

again increased from zero. For a particular reference code C_j, when the ADC output produces C_{j-1} after a complete oscillation around code $C_j(C_{j-1}-C_j-C_{j+1}-C_j-C_{j-1})$, the reference code generator code is increased by 1 LSB to generate the next reference code and make the signal generator ramp up for the next oscillation. Hence for a particular reference code C_j, the ADC is made to oscillate around it only once (to save test time) and after the first oscillation is complete, the reference code is changed to the next higher code and the ADC input ramps up for the next oscillation.

This reference code generation is performed by an up-counter which is initialized with (0000…1). During the initial delay period the D flip-flop in the reference generation circuit is disabled. Only after the first ramp-down mode has been started is the D flip-flop activated and its output becomes 'HIGH'. But only when the control signal again becomes low is the counter increased by 1 LSB.

The input to the ADC follows the profile as shown in figure 3.11.

In figure 3.12, the result for the change of reference code is shown. The first two plots show the two least significant bits of the reference code generator. It can be verified that as the control signal (last plot) becomes 'LOW' after being 'HIGH', then the reference code increases by 1 LSB.

3.3.2.3 Delay sub-block

When the input to the pipelined ADC is increased from 0 V, the corresponding digital code appears at the ADC output after the inherent delay of the ADC. Initially the signal generator output is made to be 0 V and then made to increase

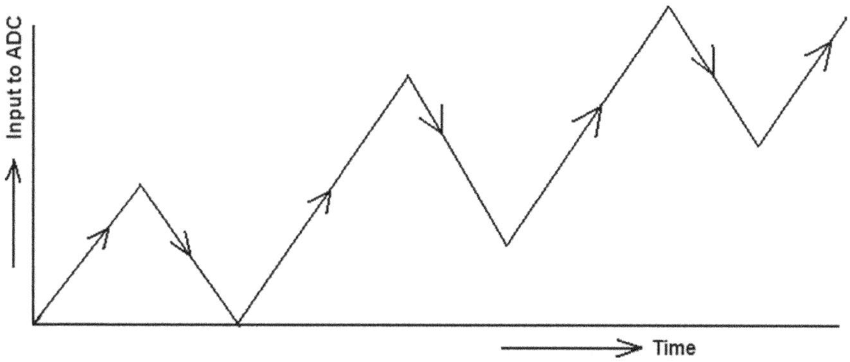

Figure 3.11. Input profile to ADC.

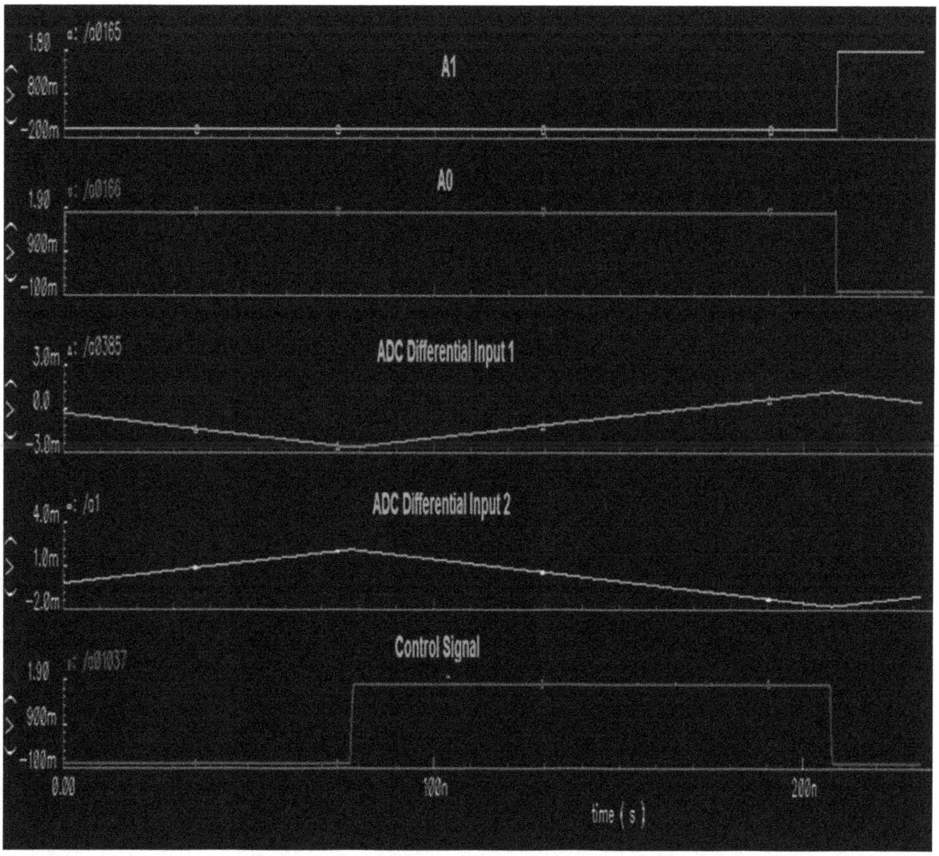

Figure 3.12. Result for a change of reference code.

linearly. Initially for the inherent delay period of the ADC, its output codes are of no interest and hence in this period the oscillation is avoided and the control signal is forcibly made to be low allowing the signal generator to increase linearly. The delay sub-block generates this delay in the circuit and stops the oscillation during that period. The delay sub-block also deactivates the error measurement blocks during the initial delay period. The pipelined ADC which has the BIST scheme has an inherent delay of eight clock cycles. In the delay block shown in figure 3.10, the D flip-flop is initialized with '0' at the output, which enables the 8 bit up-counter to count the number of clock cycles. After eight clock cycles, the D flip-flop output becomes high. This signal is used to start the oscillation. D flip-flop output also disables the counter.

3.3.2.4 Sub-block for generation of the DNL_INL_EN signal

This sub-block enables the DNL measurement block and the INL and gain error measurement block. For code-width measurement, the DNL and INL and gain measurement blocks count the number of clock cycles as long as the signal 'DNL_INL_EN' is 'HIGH'. For a particular reference code C_j, when the output of the ADC reaches C_{j+1}, the DNL_INL_EN signal becomes high. It becomes low when the code C_{j-1} appears while decreasing from C_{j+1} via C_j.

Therefore, to implement the above logic, the DNL_INL_EN signal is made to be high when the control signal is high (after the delay period) and the ADC output code is C_j. The oscillation of an ideal ADC block is shown below in figure 3.13.

The oscillation of the pipelined ADC around the code '00000001' is shown in figure 3.14.

3.3.3 Offset measurement block

Initially when the signal generator is made to ramp up from 0 V, a counter counts the number of clock cycles until the ADC produces its first code transition. Any code

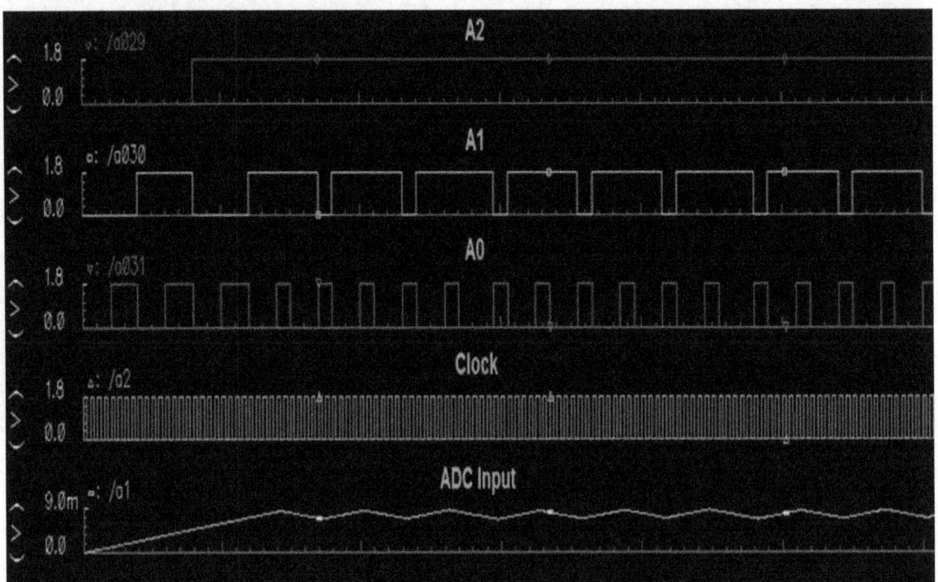

Figure 3.13. Result for oscillation of an ideal ADC around code '00000110'.

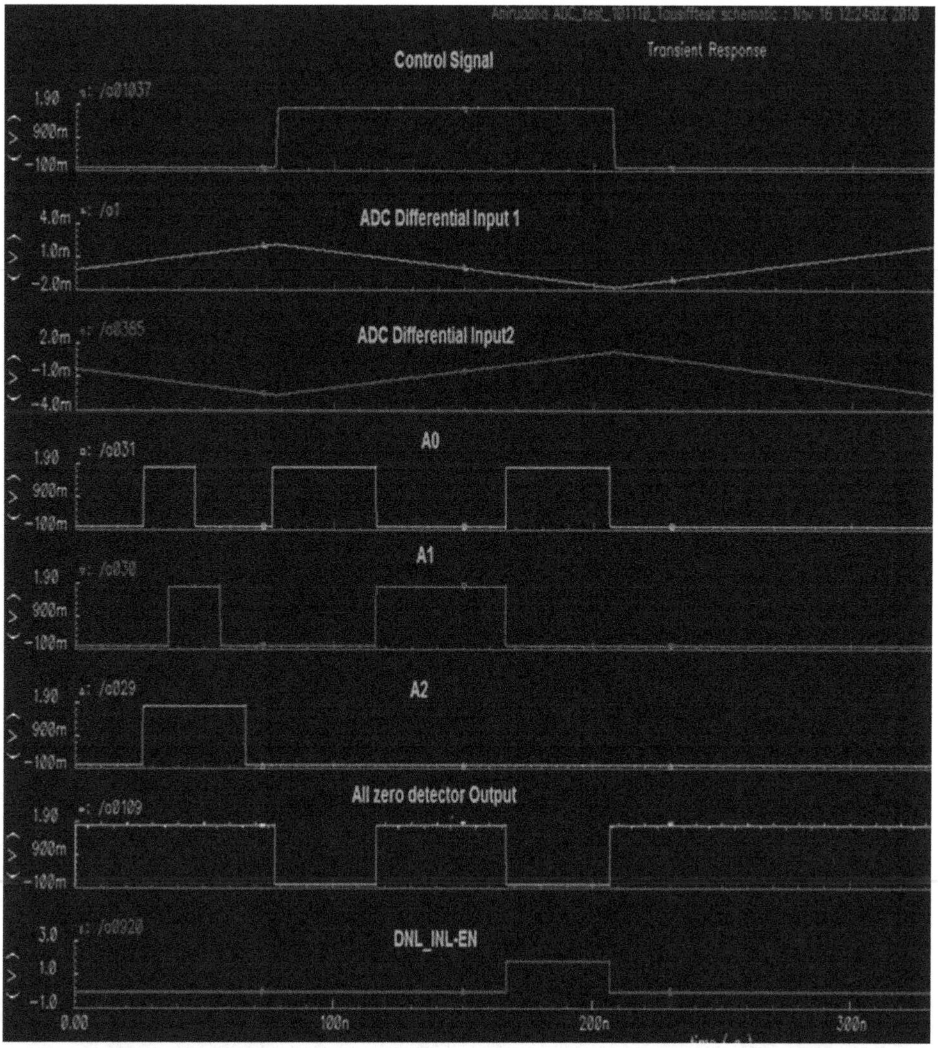

Figure 3.14. Result for oscillation of the pipelined ADC around the code '00000001'.

transition from the least significant code (000…0) can be determined by the 'OR' gate where all the bits of the ADC output code are coming as input to the gate. The counter is disabled when code transition is detected. This block is activated by the delay sub-block in the 'control logic' block after the inherent delay period of the pipelined ADC.

The implementation of the offset measurement block is shown in figure 3.15.

The simulation result for the offset measurement block is shown in figure 3.16.

3.3.4 DNL–INL–gain error measurement block

The DNL measurement block and the INL–gain error measurement block implementation are shown in figure 3.17. These blocks are also activated by the delay

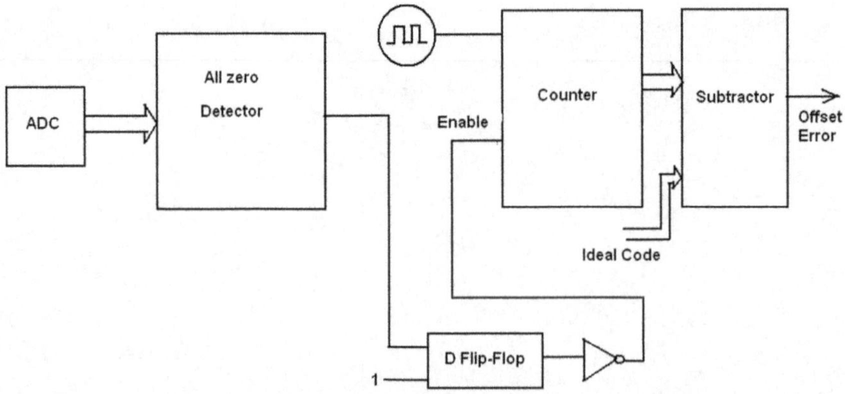

Figure 3.15. Offset measurement block.

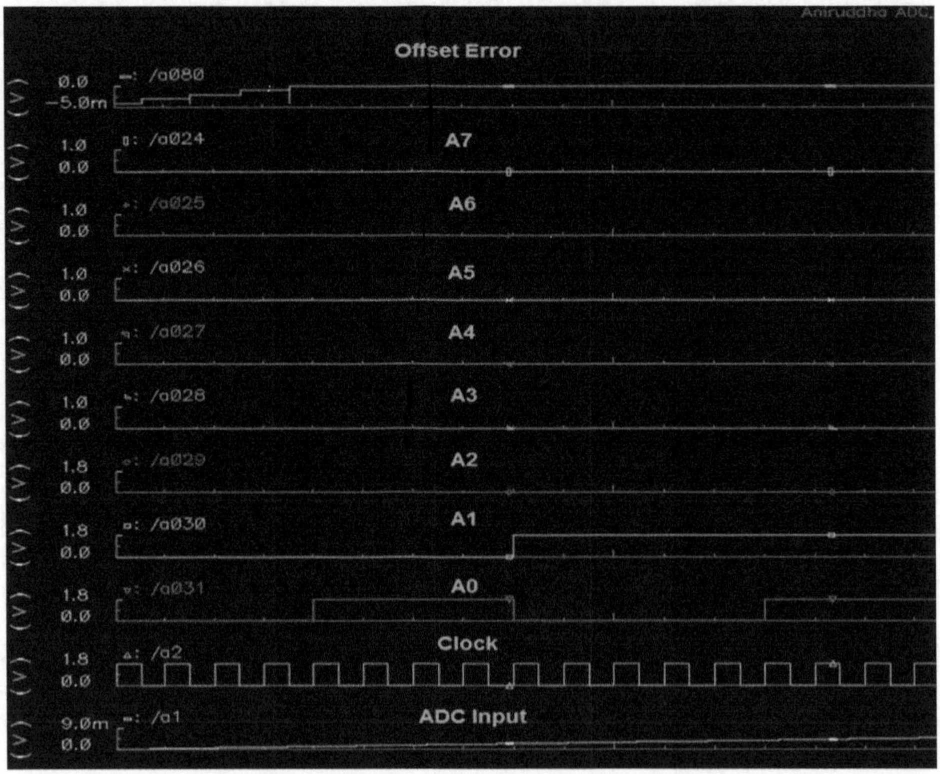

Figure 3.16. Simulation result for the offset measurement block.

sub-block in the 'control logic' block after the inherent delay period of the pipelined ADC. In the DNL measurement block, an up-counter counts the number of clock cycles during the period the ADC output decreases from C_{j+1} to C_{j-1} through code C_j. (In this period theDNL_NL_EN signal is made high by the control logic block.) The comparison of the counter output with the ideal number of clock counts gives

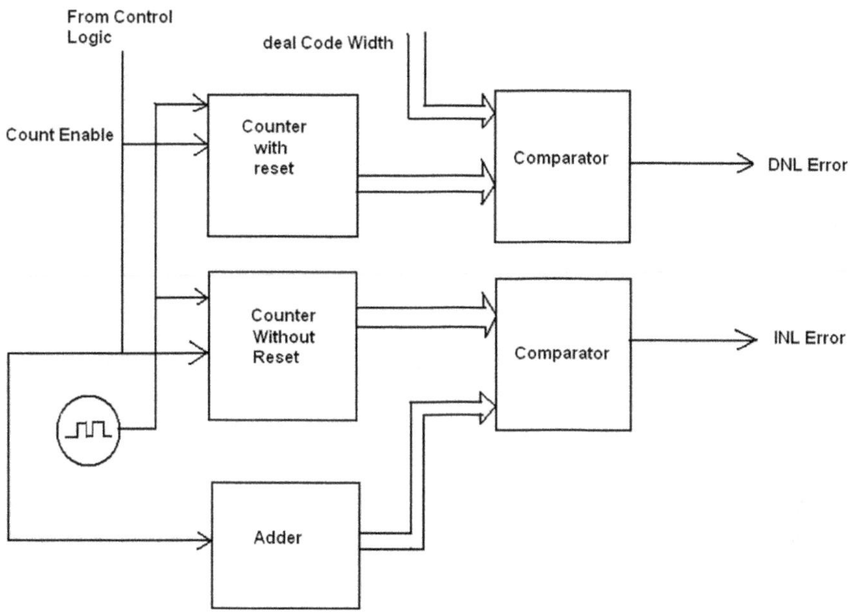

Figure 3.17. DNL–INL and gain error measurement block.

the DNL for code C_j. The counter is made to rest automatically after the DNL_INL_EN signal goes 'LOW' after being 'HIGH'.

In the INL–gain error measurement block, there is also an up-counter which counts the number of clock cycles, the same as in the DNL block. However, after the measurement period is over, the counter is not reset and hence it actually accumulates the actual code width of all the codes increasing from least significant code (000…0) to C_j. There is also an 8 bit adder which adds the ideal clock count (corresponding to ideal code width) with its output every time the counter is activated by the DNL_INL_EN signal. The comparison of the counter output with the adder output gives the INL for reference code C_j. The INL error for the reference code of the most significant code (1111…1) gives the gain error.

The simulation result for the DNL measurement block and INL–gain error measurement block are shown in figure 3.18.

3.3.5 Missing code error detection block

The missing code error detection block is shown in figure 3.19. It determines whether the difference between the consecutive codes of the ADC is higher than 1 LSB. This condition is tested when the input to the ADC is either increasing or decreasing. Here the missing code error is checked when the input to the ADC is increasing (and provided no monotonicity error is present).

The previous output of the ADC is stored in an 8 bit register. It is compared with the present ADC output. The previous output stored in the register is subtracted from the present ADC output (subtraction by two's complement method) in an 8 bit

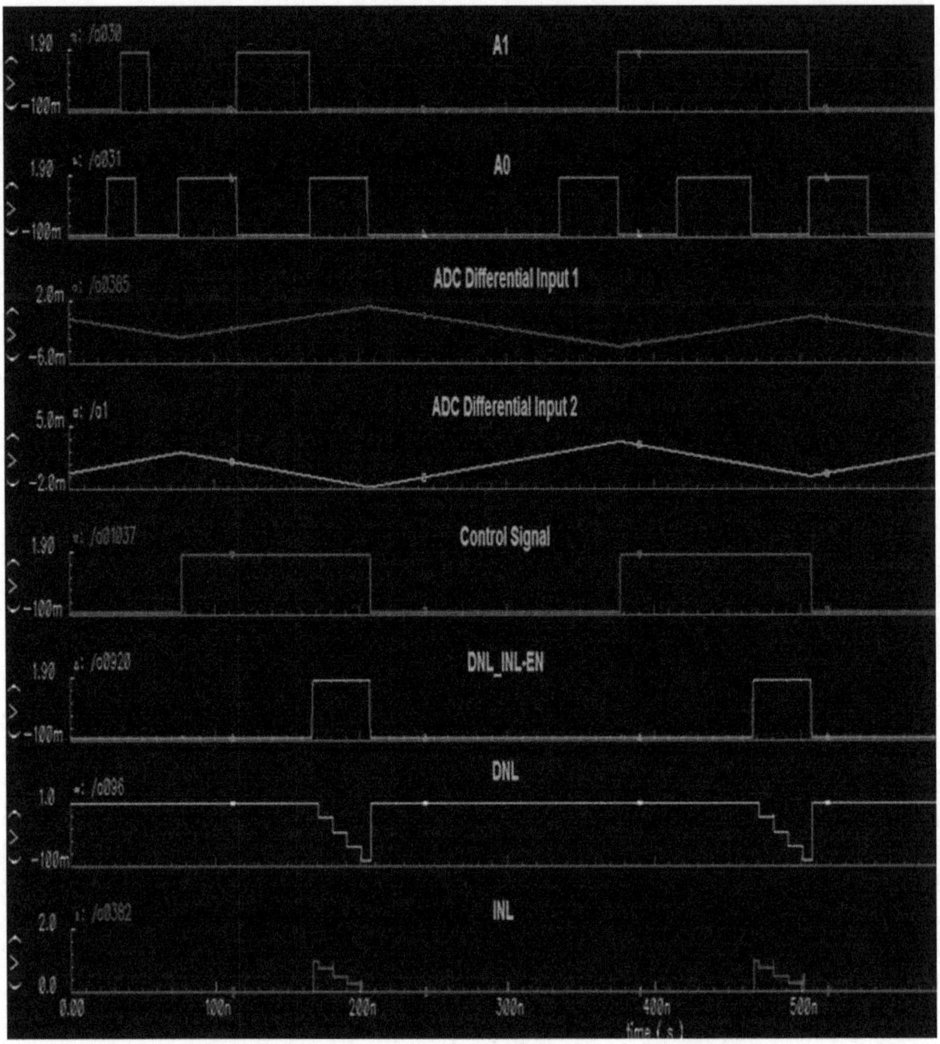

Figure 3.18. Simulation result for the DNL and INL–gain error measurement block.

subtractor. Missing code error is determined when the input to the ADC is increasing. If any one of the seven output bits (LSB is excluded) of the 8 bit subtractor output is high then there is missing code error in the ADC. This is achieved by simply 'OR'ing all seven bits. This block is also activated after the initial delay period of the ADC is over (eight clock cycles).

3.3.6 Monotonicity error detection block

The monotonicity error detection block determines whether there is a decrease in the ADC output code, when the ADC input is increased monotonically. The oscillation-based BIST fails when there is monotonicity error in the ADC. Hence to test it, the oscillation is disabled and the input to the ADC is increased monotonically. This is

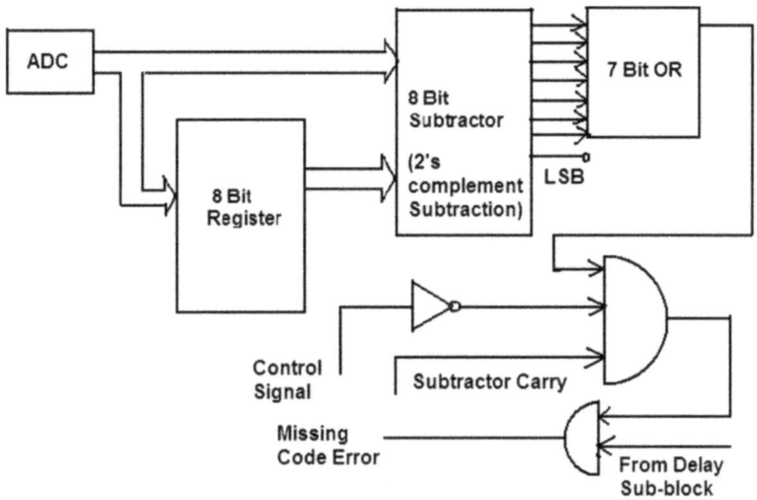

Figure 3.19. Missing code error detection block.

done by a select line named 'oscillation disable'. The high or low status of the select line decides whether to make the oscillation and measure the offset error, DNL, INL, gain error and missing code error, or to stop the oscillation and detect the monotonicity error. As in the missing code detection block, the previous output of the ADC is stored in an 8 bit register. It is compared with the present ADC output. The previous output stored in the register is subtracted from the present ADC output (subtraction by the two's complement method) in an 8 bit subtractor. If the generated carry is high, then the present output is higher than the previous output and there is no monotonicity error. But a 'LOW' carry indicates that the present output is lower than the previous output and hence the ADC is faulty with monotonicity error. It is important to note that we cannot test for monotonicity error and missing code error simultaneously. Hence the 8 bit register and the 8 bit subtractor used in the missing code error detection block can also be shared with the monotonicity detection block. It will further reduce the silicon area.

The implementation of the monotonicity error detection block is shown in figure 3.20.

3.4 Introduction to ADC dynamic testing

3.4.1 Introduction

The dynamic specifications of ADCs are very important for high speed communication applications such as digital communication, ultrasound imaging, instrumentation and intermediate frequency digitization. It is more convenient to characterize the performance parameters in the frequency domain. Data converters react differently when a frequently changing input is applied to them. Hence dynamic testing mainly deals with application of a high frequency input that is a single-frequency sinusoid signal to the ADC and test how the ADC behaves under it. The dynamic

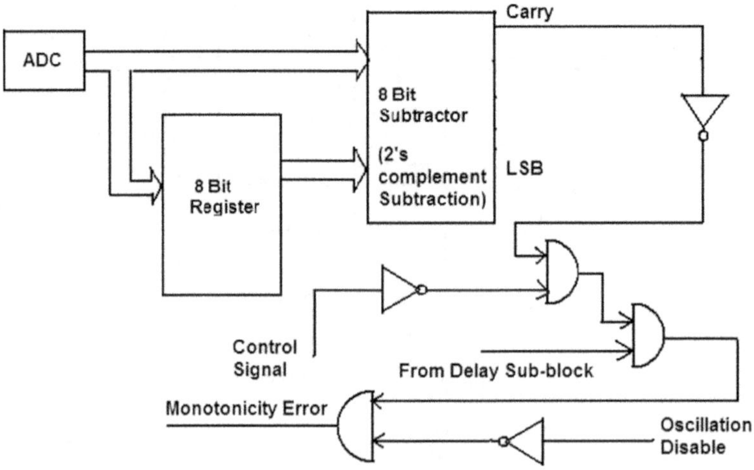

Figure 3.20. Monotonicity error detection block implementation.

specification of ADCs involves the SNR, effective number of bits (ENOB), total harmonic distortion (THD), spurious free dynamic range (SFDR), etc. Now the dynamic testing requires a spectrally pure high frequency sine wave as an input to the ADC. In this work a spectrally pure analog sine wave is created using oversampling digital-to-analog conversion methods.

The design constraints for a high precision, analog signal source for BIST applications are mainly:

1. The source should be capable of generating high precision test signals with the quality exceeding that of the CUT.
2. The source should be insensitive to parameter variation.
3. The circuit must occupy minimal silicon area.

Conventional analog sinusoid generation principles such as the Wien bridge oscillator, RC phase shift oscillator, tuned circuit oscillator, etc, already exist, but they require bulky circuit components and also the frequency of oscillation is not easily adjustable. Furthermore, the analog oscillator design will be vulnerable to variations in the fabrication process and temperature drift. In this work a sine generation system using digital techniques is implemented. Then an oversampling digital-to-analog conversion technique is used to generate an analog sinusoid from the digitally generated sinusoid.

3.4.2 Lossless discrete integrator

Lossless discrete integrator is the digital version of the analog LC tank circuit which is lossless. Being a lossless circuit it follows that once the LC tank is excited, no energy is lost but instead alternates between electric and magnetic forms.

The resulting capacitor voltage and inductor current waveforms represent ideal sinusoids with frequency $\omega_0 = \frac{1}{\sqrt{LC}}$. Figure 3.21 shows the parallel LC tank circuit.

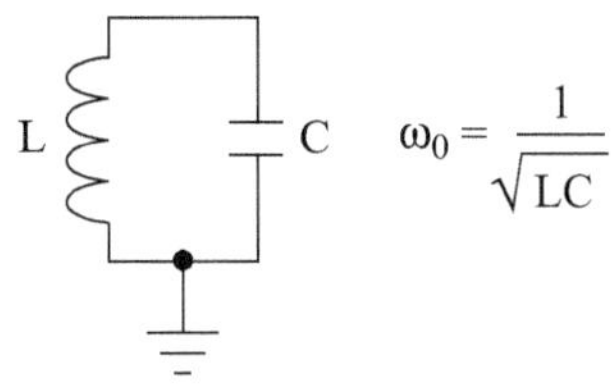

Figure 3.21. Parallel LC tank circuit.

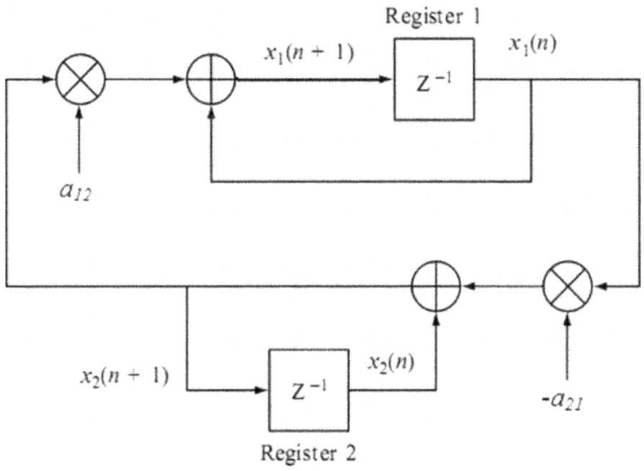

Figure 3.22. Second-order digital resonator.

The unique feature of the LC tank circuit is that the variations in the capacitor or inductor value do not prevent the circuit from oscillating, but merely change the frequency of oscillation. Second, the amplitude of the oscillation is a function of the initial conditions imposed on the capacitor and inductor.

The digital equivalent of the LC tank circuit is shown in figure 3.22. Two discrete time integrators in a loop form the resonator circuit. They represent the two differential equations of the LC tank circuit.

The two difference equations characterizing the oscillator are

$$x_1(n + 1) = x_1(n) + \alpha_{12}x_2(n + 1) \tag{3.7}$$

$$x_2(n + 1) = -\alpha_{21}x_1(n) + x_2(n). \tag{3.8}$$

From these two equations, $x_2(n)$ can be eliminated to give a single equation in $x_1(n)$. Taking the Z transform of the resultant equation, with $X_1(z)$ representing the z transform of $x_1(n)$, one obtains

$$z^2X_1(z) + (\alpha_{12}\alpha_{21} - 2)zX_1(z) + X_1(z) = 0. \tag{3.9}$$

The characteristic equation is given by

$$z^2 + (\alpha_{12}\alpha_{21} - 2)z + 1 = 0. \tag{3.10}$$

The roots of equation (3.10) lie on the unit circle for all values of the product $a_{12}a_{21}$ between 0 and 4. Hence oscillation is guaranteed for this constraint. If the initial condition of the variables $x_1(n)$ and $x_2(n)$ are $x_1(0)$ and $x_2(0)$, respectively, then the frequency (rad/sec), amplitude and phase of the generated sinusoidal signal are given as follows:

Frequency (rad/sec)

For $0 < \alpha_{12}\alpha_{21} \leqslant 2$,

$$\omega_0 = f_{\text{ck}} \cos^{-1}\left(1 - \frac{\alpha_{12}\alpha_{21}}{2}\right). \tag{3.11}$$

For $2 < \alpha_{12}\alpha_{21} \leqslant 4$,

$$\omega_0 = f_{\text{ck}}\left[\pi - \cos^{-1}\left(1 - \frac{\alpha_{12}\alpha_{21}}{2}\right)\right]. \tag{3.12}$$

Amplitude

$$A = \frac{(1 - \alpha_{12}\alpha_{21})x_1(0) + \alpha_{12}x_2(0)}{\sin(\omega_0 T + \Phi)}. \tag{3.13}$$

Phase

$$\Phi = \tan^{-1}\left(\frac{x_1(0)\sin(\omega_0 T)}{(1 - \alpha_{12}\alpha_{21} - \cos(\omega_0 T))x_1(0) + \alpha_{12}x_2(0)}\right). \tag{3.14}$$

Here $f_{\text{ck}} = \frac{1}{T}$ is the clock frequency of the resonator circuit. As was stated earlier, it can be verified that the oscillation frequency is dependent only on $\alpha_{12}\alpha_{21}$ and, if due to parameter variation $\alpha_{12}\alpha_{21}$ changes, it will not prevent the circuit from oscillation, but just changes the frequency of oscillation. Second, the amplitude of the oscillation will be a function of the initial conditions imposed on the two registers.

3.4.3 Analog sine generator

The analog sine can be generated from the digital oscillator by connecting an N-bit DAC or a 1 bit oversampling DAC at the register output. But these two options require a large silicon area. The 1 bit oversampling DAC uses an interpolation filter which requires a large silicon area.

The alternative design option alleviates this problem by operating the entire resonator loop at the oversampling rate and hence eliminating the need for an interpolation filter. Furthermore, if we insert the delta-sigma modulator of the 1 bit oversampling DAC inside the resonator loop, the multi-bit multiplication blocks can be replaced very efficiently, saving a large area of silicon [23, 24].

3.4.4 Simulated results

The signal generator circuit is simulated in MATLAB and the simulated results are given in figures 3.23 and 3.24.

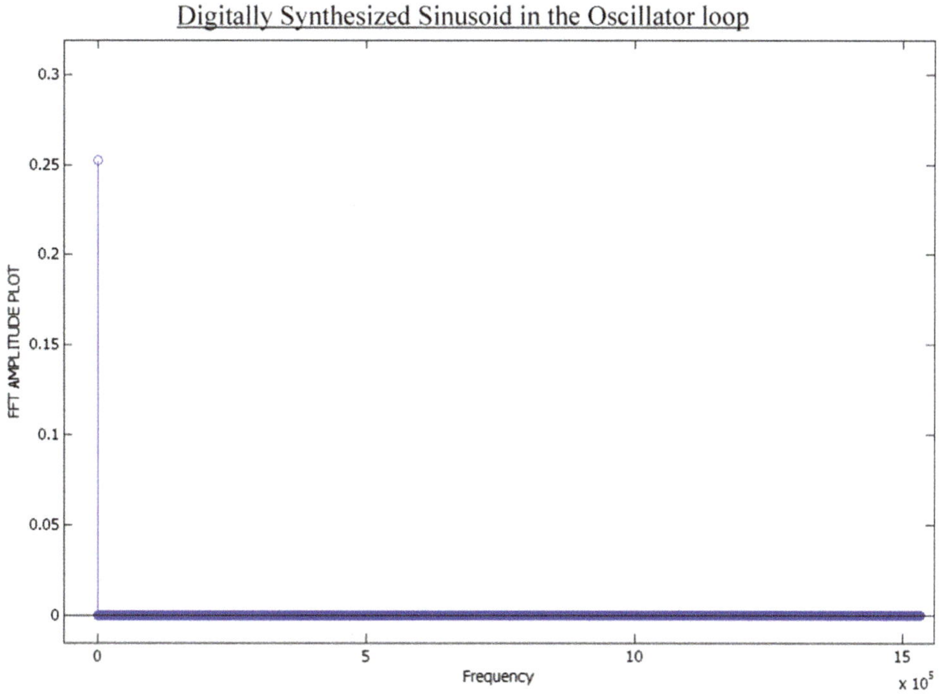

Figure 3.23. FFT magnitude plot.

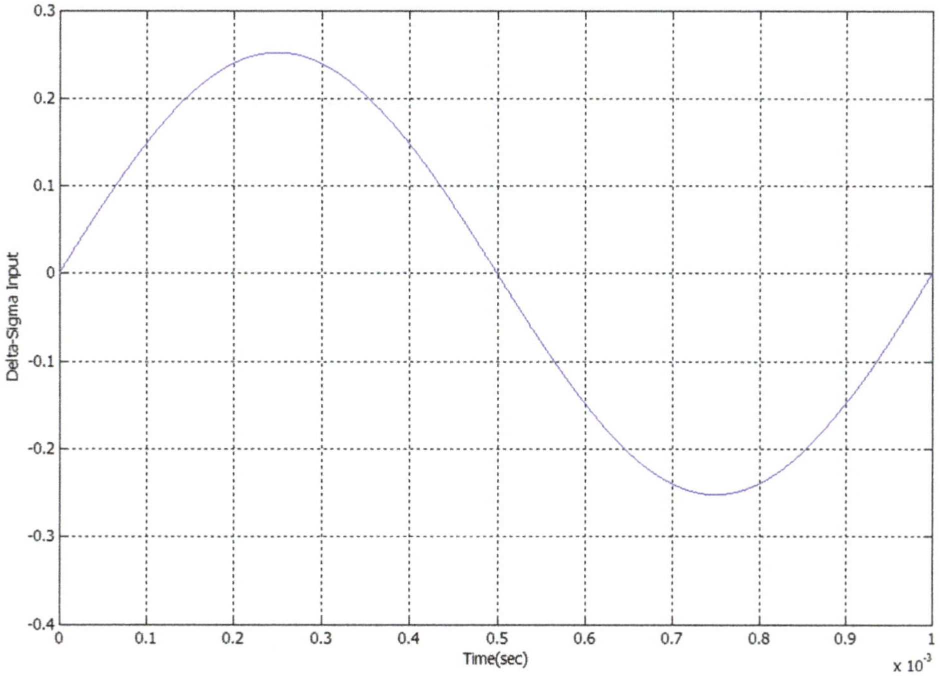

Figure 3.24. Time domain plot.

3.5 Conclusion

3.5.1 Work performed

In this work a novel oscillation based on a BIST system has been designed for a 1.8 V, 8 bit pipelined ADC in 180 nm standard CMOS technology in the Cadence environment. Oscillation-based BIST is a general BIST scheme applicable to analog and mixed signal circuits. This technique has been previously applied on analog filters and operational amplifiers. In this work this test technique is applied for the functional testing of a pipelined ADC. The ADC is reconfigured to oscillate around a particular output code and from the code-width measurement the static performance parameters, namely offset error, differential non-linearity (DNL), integral non-linearity (INL), gain error, missing code error and monotonicity error, are measured. The FNC, has been avoided, which saves extra chip area overhead. A ramp generator has been built which ramps up or ramps down according to a 'control signal' status. The 'control logic' block controls the entire OBIST operation and controls the error measurement and detection blocks. The control strategy for oscillation has been modified to take care of the latency requirement of the pipelined ADC. Four sub-blocks are made inside the 'control logic' block to control its different functionalities. The control signal generation sub-block starts and maintains the oscillation of the ADC by controlling the signal generator. The reference code generation sub-block generates reference codes sequentially for which the linearity errors will be measured. To take care of the pipelined ADC inherent delay period, delay sub-block disables the oscillation for the initial delay period. It also deactivates the error measurement blocks for this period. The enable signal for DNL and INL–gain error measurement blocks is generated by the DNL_INL_EN signal generation sub-block. The offset measurement block, DNL measurement block, INL and gain error measurement block, missing code error detection block and monotonicity error detection blocks are designed separately to measure the respective static performance parameters of the ADC.

For testing the dynamic performance parameters of the ADC an oversampling-based sinusoidal generator has been simulated and tested in MATLAB. The generator has been simulated in the Cadence environment, also in 180 nm standard technology. Instead of sticking to conventional analog sine generation principles, an LDI-based digital sine generator has been developed, which is followed by a DAC and a low-pass filter.

3.5.2 Future scope of work

1. The chip layout for the pipelined ADC and the associated oscillation-based BIST circuit can be created. Then the chip can be fabricated and tested for the desired results.
2. A BIST system can be built for dynamic performance testing of the designed ADC.

References

[1] Stroud C E 2002 *A Designer's Guide to Built-In-Self-Test* (Dordrecht: Kluwer)

[2] Doernberg J, Lee H-S and Hodges D A 1984 Full-speed testing of A/D converters *IEEE J. Solid State Circuits* **SC-19** 820–7

[3] Mahoney M 1987 *DSP-based Testing of Analog and Mixed-Signal Integrated Circuits* (New York: IEEE Computer Society Press)

[4] Renovell M, Azaïs F, Bernard S and Bertrand Y 2000 Hardware resource minimization for a histogram-based ADC BIST *Proc. 18th IEEE VLSI Test Symp. (Montreal, Canada, April–May 2000)* pp 1–6

[5] Azaïs F, Bernard S, Betrand Y and Renove M 2000 Towards an ADC BIST scheme using the histogram test technique *Proc. IEEE European Test Workshop (Cascais, Portugal, May 2000)* pp 53–8

[6] Azaïs F, Bernard S, Bertrand Y and Renovell M 2001 Implementation of a linear histogram BIST for ADCs *Proc. Design, Automation and Test in Europe, Conf. and Exhibition (Munich, Germany, March 2001)* pp 590–5

[7] de Vries R, Zwemstra T, Bruls E M J G and Regtien P P L 1997 Built-in self-test methodology for A/D converters*(Proc. European Design and Test Conf. (ED & TC)* (Paris, France, March 1997) *pp 353–8*

[8] Lee D, Yoo K, Kim K, Han G and Kang S 2004 Code-width testing-based compact ADC BIST circuit *IEEE Trans. Circuits Systems-II, Express Briefs* **51** 603–6

[9] Ehsanian M, Kaminska B and Arabi K 1996 A new digital test approach for analog-to-digital converter testing *Proc. of 14th VLSI Test Symp. (Princeton, NJ, April–May 1996)* pp 60–5

[10] Huang J-L, Ong C-K and Cheng K-T 2000 A BIST scheme for on-chip ADC and DAC testing *Proc. Design, Automation and Test in Europe Conf. and Exhibition (Paris, March 2000)* pp 216–20

[11] Erdogan E S and Ozev S 2007 An ADC-BIST scheme using sequential code analysis *Proc. of Design, Automation & Test in Europe Conf. & Exhibition (Nice, France, April 2007)* pp 1–6

[12] Da Gloria Flores M, Negreiros M, Carro L, Susin A A, Clayton F R and Benevento C 2005 Low cost BIST for static and dynamic testing of ADCs *J. Electron. Test.* **21** 283–90

[13] Duan J, Jin L and Chen D 2010 INL based dynamic performance estimation for ADC BIST *Proc. of IEEE Int. Symp. On Circuits and Systems (ISCAS) (May–June 2010)* pp 3028–31

[14] Arabi K and Kaminska B 1997 Efficient and accurate testing of analog-to-digital converters using oscillation-test method *Proc. European Design and Test Conf. (ED & TC) (Paris, France, March 1997)* pp 348–52

[15] Wang Y-S, Zhang J-L, Yu M-Y and Xiao L-Y 2007 A novel oscillation-based BIST for ADCs *7th Int. Conf. on ASIC (Guilin, China, October 2007)* pp 1010–3

[16] Nicholas H T and Samueli H 1991 A 150-MHz direct digital frequency synthesizer in 1.25-μm CMOS with −90-dBc spurious performance *IEEE J. Solid State Circuits* **26** 1959–69

[17] Nicholas H T, Samueli H and Kim B 1988 The optimization of direct digital frequency synthesizer performance in the presence of finite word length effects *Proc. of 42nd Annual Frequency Control Symp. (Baltimore, MD, June 1988)* pp 357–63

[18] Lu A K, Roberts G W and Johns D A 1993 High quality analog oscillator using oversampling D/A conversion techniques *Proc. of ISCAS '93 (Chicago, IL, May 1993)* pp 1298–301

[19] Lu A K, Roberts G W and Johns D A 1994 A high-quality analog oscillator using oversampling D/A conversion techniques *IEEE Trans. Circuits Systems-II: Analog Digit. Signal Process.* **41** 437–44

[20] Allen P E and Holberg D R 2002 *CMOS Analog Circuit Design* (Oxford: Oxford University Press)

[21] Mallikarjun E 2009 A 1.8 V 10-bit 500 Ms/s parallel pipelined ADC *MTech Dissertation* IIT Kharagpur, Department of Electrical Engineering

[22] Tausiff D M D 2010 A built-in self test on 1.8 V 8-bit 125 MSPS pipelined ADC *MTech Dissertation* IIT Kharagpur, Department of Electrical Engineering

[23] Roberts G W and Lu A K 1995 *Analog Signal Generation for Built-In-Self-Test of Mixed-Signal Integrated Circuits* (Berlin: Springer)

[24] Kabisatpathy P, Barua A and Sinha S 2005 *Fault Diagnosis of Analog Integrated Circuits* (Dordrecht: Springer)

Chapter 4

An oscillation-based built-in self-test (BIST) system for dynamic performance parameter evaluation of an 8 bit, 100 MSPS pipelined ADC

Nalla Dhanunjay and Alok Barua

An oscillation-based built-in self-test (BIST) scheme is proposed for dynamic parameter testing of a pipelined analog-to-digital converter (ADC). In the BIST structure, an ac (1 V p–p, 1 KHz–1 MHz) signal has been applied to the pipelined ADC to find the dynamic parameters and the output code is passed through a digital-to-analog converter (DAC) algorithm (in simulation) to reconstruct the input signal. A 1024 point fast Fourier transform (FFT) has been performed on the reconstructed signal. From the FFT plot/data the dynamic parameters have been calculated. To generate an on-chip ac signal, a filter-based approach is presented. A square wave with variable frequency is applied to the second-order low-pass filter through a wave shaping circuit which transforms the square wave into a harmonic-free digital sine wave. The filter is used for removing the unnecessary higher harmonics and noise components. The signal obtained from the filter is fed to the ADC and the dynamic performance parameters are evaluated from the FFT plot of the reconstructed signal.

By using the BIST system some of the most likely faults, such as offset errors of the comparators, gain errors of the OTAs and the effect of capacitor mismatch, have been introduced in the pipelined ADC, and also the effect of the faults on the dynamic performance parameters of the system have been tested successfully.

4.1 Introduction

4.1.1 Introduction to BIST

Testing is a very important part in the post-manufacturing process of a very large-scale integration (VLSI) chip. During the design phase, testing is performed to detect and identify design faults, while in the manufacturing phase testing is performed to

doi:10.1088/978-0-7503-1732-0ch4

avoid manufacturing defects. Finally, during system operation testing seeks to detect any fault sustained during operation that would produce erroneous output from the system. To ensure that a fault-free product reaches the customer, rigorous testing methods are adopted during the product lifecycle. However, the cost of testing and the test time are the two major factors that influence the stringent test requirements to be satisfied and they definitely increase the cost of the product. One good alternative is to have a BIST system embedded in the VLSI device.

The increasing functional complexity of electronic systems makes testing a challenging task to yield fault-free devices or systems. Considering that testing represents a key cost factor in the production process (a proportion of up to 70% of total product cost as reported in [1]), an optimal test strategy can be a substantial competitive advantage in a system comprising a large number of transistors.

Some of the early approaches using BIST to create input stimulus for testing of the analog portion of mixed-signal systems were to make use of digital test pattern generator (TPG) and output response analyser (ORA) functions that were typically used for testing digital circuitry. TPG functions are produced by linear feedback shift registers (LFSRs) which generate pseudorandom digital patterns that look similar to white noise when passed through a DAC. However, ramp input signals have been used in analog testing and were found to provide good fault detection and, in some cases, better results than sinusoidal test signals. In addition, it has been observed that the detection of faults with respect to the input test waveform can vary with the type of analog circuit under test. Therefore, a variety of different test waveforms are needed to provide good fault detection coverage in a wide range of analog application circuits.

The BIST for digital circuits has been an active area of research and development for more than two and a half decades and there are a number of good BIST approaches to test the digital portion of mixed-signal systems. However, BISTs for analog circuitry have received much less attention until recently. As a result, testing the analog portion of mixed-signal integrated circuits and systems has been identified as one of the major challenges for the future and BIST has been identified as one of the good solutions for this testing procedure.

Generally there are two types of testing methods, one is functional testing and the other one is structural testing. Functional testing verifies the functional working of a particular circuit/system, whereas structural testing involves testing the physical structure, to verify the physical operation.

ADCs are a very important component of mixed-signal circuits. Efforts are being made to develop new ADC designs which will ensure high sampling speed, high accuracy and low power. Hence new and efficient test algorithms should also be developed to test high performance ADCs. In this work a new BIST system has been developed for these high performance ADCs with different architectures.

The testing of ADCs can be classified into static testing and dynamic testing. In static testing a slowly varying (low frequency) signal is applied and the following test parameters are evaluated from its output response.

1. Offset error.
2. Differential non-linearity (DNL).

3. Integral non-linearity (INL).
4. Gain error.
5. Missing code error.
6. Monotonicity error.

For dynamic testing of ADCs a high frequency sinusoidal signal is applied to ADC and from the output response the following dynamic performance parameters are evaluated:

1. Signal-to-noise ratio (SNR).
2. Total harmonic distortion (THD).
3. Signal-to-noise plus distortion ratio (SINAD).
4. Spurious free dynamic range (SFDR).
5. Effective number of bits (ENOB).

In this work an oscillation-based BIST system is designed for the dynamic testing of a 1.8 V, 8 bit, 100 MSPS pipeline ADC. In this test scheme a 1 V peak-to-peak and a frequency of range 1 KHz to 1 MHz is applied to the ADC under test. The output code is reconstructed using a DAC algorithm to make it more precise. A 1024 point fast Fourier transform (FFT) is applied to the reconstructed signal to analyse the dynamic performance parameter.

To generate the test stimulus a new scheme is proposed to convert a square wave into a sine wave, with a total harmonic distortion (THD) of 0.00037%. This is achieved with the help of the new scheme, which makes the square wave, a summation of all the odd harmonics, into a harmonic-free stepped sine wave, which contains only $8n \pm 1$th harmonics. After obtaining the stepped sine wave it is passed through an operational transconductance amplifier and capacitor (OTA-C) filter to eliminate the existing $8n \pm 1$th harmonics.

The OTA-C filter is the better trade-off between the performance and the area of silicon occupied by it. Before using the OTA-C filter the standard filter structures, such as the active RC filter and switched capacitor filter, were tried. The active RC filter gives a much better performance but with a higher silicon area, and the switched capacitor filter has a lower area overhead but its performance is poor, i.e. instead of filtering it introduces harmonics and noise in the signal output.

As the OTA-C filter gives higher performance with a smaller area overhead, it requires some tuning in order to avoid its performance being affected by changes in process variations, supply voltages and temperatures. The on-chip digital automatic tuning procedure is presented to tune the filter parameters for better performance, and because of the digital elements employed in the tuning procedure the silicon occupied by the tuning system is minimal.

After successful generation of a test stimulus, it is applied to the ADC and the dynamic performance parameters are evaluated from the ADC response for that test stimulus. The output response of the ADC is reconstructed with the help of a DAC, which is designed with the Verilog-A code with 2 bits more resolution than the ADC and the FFT is applied on the reconstructed signal to evaluate the dynamic performance parameters for the healthy pipelined ADC.

Later, some of the most common faults in ADCs are created manually to study the behaviour of the ADC for these particular faults, and the dynamic performance parameters for the faulty system are evaluated and compared with the healthy system's parameters.

Finally, the layout for the entire system, including the circuit under test and its associated testing circuit, is prepared, and the total chip area is 0.1666 mm^2, where the cut has an area of 0.1271 mm^2, which is about 76.3% of the total chip area, i.e. the testing circuit occupies only 23.7% of the total chip area.

4.1.2 Literature review

Several BIST techniques for dynamic testing of ADCs have been published in recent papers. The requirements for testing circuits working on the principle of a self-system are discussed by Kang [1] and Rueda [2]. Novac proposed a method for measuring the dynamic performance parameters and the accuracy of the testing system [3]. Savino discusses the performance of FFTs for measuring the dynamic parameters [4]. The basics of integrated circuits and the basic building blocks for analog integrated circuits, from the fundamentals with the use of metal oxide semiconductor field effect transistors, are presented in [5] and [6].

The test stimulus generation techniques using sine waves, and their advantages and disadvantages, are presented in [2] and [7]. The special technique for generating a test stimulus is presented in [8]. The analog filter, the discrete time filters and the transconductance–capacitor filters are covered in detail in [9] and [10]. The transconductance capacitor filter is emphasised in [11], while [12] discusses the very small transconductance filters and achieving transconductance in the nano-ampere per volts range. The g_m–C filters and OTA-based circuits are discussed in other publications, e.g. [13]. The tuning methods for filters and CMOS op-amps are discussed in [14] and [15]. The design procedure for pipelined ADCs is discussed in [16, 17] and [18]. Savino explains an FFT test of ADC in [4]. The frequency domain analysis for measuring the dynamic performance parameters of the ADCs and also for measuring the INL and DNL from the frequency response of the ADC was presented by Petri *et al* [19]. A self-testing scheme for the pipelined ADC is discussed, along with the calibration techniques for improving the performance parameters, in [20].

The testing schemes for the faults in the pipelined ADC, and their effect on the system performance, are presented in [21, 22]. The capacitor mismatch error and its calibration procedure are presented by Hamoui [23]. The gain error calibration technique is discussed in [24]. The different kinds of sources of noise in the pipelined ADC are presented by Galton [25]. Designing pipelined ADC and BIST systems for measuring different parameters of ADC was reported by workers at IIT Kharagpur [26–30].

4.1.3 Motivation and aims

The oscillation-based BIST method is a general test scheme applied to functional and structural testing of mixed-signal circuits. The basic principle of this test scheme is that the circuit under test is reconfigured to oscillate and the oscillation frequency of the system is to be observed. The deviation of the frequency of oscillation from its nominal value indicates a faulty circuit. The oscillation-based test strategy has been successfully applied to a wide range of analog and mixed-signal circuits, including analog filters and operational amplifiers. However, the application of this test strategy has never been tested for the functional testing of pipelined ADCs. To generate the test stimulus, in most cases an oscillator is used, but the square wave to sine wave converters have not been used previously, because of the practical limitations of the filters and their order.

With the motivation above, the goal of this work is to apply the oscillation-based BIST methodology for functional testing of pipelined ADCs. A 1.8 V, 8 bit, 100 MSPS pipelined ADC is used as the circuit under test (CUT) to test the OBIST methodology and a new scheme is used to subsequently convert the square wave into a sine wave to evaluate the dynamic performance parameters listed below:

1. Signal-to-noise ratio (SNR).
2. Total harmonic distortion (THD).
3. Signal-to-noise plus distortion ratio (SINAD).
4. Spurious free dynamic range (SFDR).
5. Effective number of bits (ENOB).

4.1.4 Contributions of this chapter

An oscillation-based BIST system is developed for a 1.8 V, 8 bit, 100 MSPS pipelined ADC, which was designed by previous workers. The on-chip test circuit is designed and simulated in the Cadence platform with standard 180 nm technology. Initially the test stimulus is generated using a square wave to harmonic-free stepped sine wave converter, which consists of simple delay elements and amplifiers. The active RC filter and switched capacitor filters are designed to remove the unnecessary harmonic components which still remain after processing through the signal processing block to remove the harmonic components. Unfortunately, these filters do not serve the intended purpose, as they either need a large area on silicon or their performance is not good. To overcome these difficulties the continuous time operational transconductance amplifier and capacitor (OTA-C) filter has been developed.

The performance of the filter depends on the linearity issues of the transconductor amplifier, and hence the OTA design is selected such that its linearity is high enough. Because of the smaller dimensions of the transistors used for the trans-conductor amplifiers, they are prone to parasitic effects. The on-chip automatic tuning system for the OTA-C filter is presented. The dynamic performance parameters for a healthy or fault-free ADC are calculated from the output response for the test stimulus.

Some of the faults are created in the ADC and are tested with the help of a BIST system. The behaviour of these faults is studied, and the dynamic performance parameters are once again calculated for the faulty system. To ensure that the proposed test stimulus generator and the associated testing circuits have the minimum possible silicon area, the layout for the entire system, including the BIST and CUT, is prepared.

4.1.5 Chapter organization

This chapter is organized as follows.

Section 4.2 discusses the basic structure of the BIST scheme. The general principles for oscillation-based BIST systems are presented. Then the oscillation-based BIST principles and BIST structure for functional testing of ADCs are discussed in detail. In section 4.3, the test stimulus generation is discussed. How most on-chip test stimuli are generated and their advantages and limitations are discussed. A new technique to eliminate the harmonic contents of the square wave signal is presented, and the practical circuit implementation of the same, along with all of the blocks required for this, are also discussed. The responses of the analog and discrete time filters to the four-level sine wave are studied. The advantages and limitations of the analog filters in terms of stimulus generation are presented. In section 4.4, a detailed analysis of the OTA-C filter is presented. The basic requirements for the OTA-C filter and the design of an OTA with higher linearity using some special techniques is presented. This section discusses the on-chip automatic tuning system for the OTA-C filter. The requirements for the filter tuning circuits and their working principles are presented. The digital technique for tuning the filter is discussed. In section 4.5, the basics of the pipelined ADC are discussed and the dynamic performance parameter evaluation is presented for the same, by applying the test stimulus generated from the OTA-C filter in the OBIST system. In section 4.6, the various sources of error such as switching charge injection, thermal noise and the non-idealities of the important building blocks, and their effect on the system and on the dynamic performance parameters of the system, are presented. The results for some of the faults are also discussed. Section 4.7, presents the details of the layout preparation for the BIST system and the post-layout simulations for the same. Section 4.8 presents our conclusions and briefly explains the work done to prepare this chapter.

4.2 Oscillation-based BIST principles

4.2.1 General BIST principles

The commercial trends of the integrated circuit (IC) industry, including telecommunications, multimedia, instrumentation, automotive, etc, have forced the realization of complex mixed-signal electronic systems consisting of densely integrated analog and digital circuitry on a single IC substrate. Consequently, the increasing cost associated with testing and fault diagnosis of these complex systems has motivated research efforts to explore efficient testing methodologies. This issue is identified in the SIA roadmap for semiconductors [1] as one of the key problems for current and future mixed-signal systems-on-chips (SOCs).

Usually, the main test difficulties are due to the testing of the analog parts. Traditional test methods for analog circuits rely on functional tests, but the sensitivity of analog cores to loading effects, environmental conditions and process variations makes their testing a difficult task. Moreover, they demand high quality input stimuli, and high data volume acquisition and processing capability, etc, requiring expensive automatic test equipment (ATE). BIST schemes are a well-accepted solution to overcome some of these problems. These schemes consist of moving part of the required test resources from the ATE to the chip.

A system is tested and diagnosed on numerous occasions during its lifetime. It is critical to have quick and very high fault coverage testing. One common and widely used method in the semiconductor industry for IC chip testing is to specify the test as one of the system functions and thus it becomes a self-test. A system designed without an integrated test strategy which covers all levels from the entire system to components is described as chip-wise and system-foolish. A properly designed BIST is able to offset the cost of added test hardware while at the same time ensuring the reliability, testability and reduced maintenance costs of the system.

The basic idea of the BIST, in its most simple form, is to design a circuit so that it can test itself and determine whether it is 'good' or 'bad' (fault-free or faulty, respectively). This typically requires additional circuitry whose functionality must be capable of generating test patterns as well as providing a mechanism to determine if the output responses of the circuit under test (CUT) to the test patterns correspond to that of a fault-free circuit.

4.2.2 BIST basic test flow

The basic test flow of a BIST is shown in figure 4.1.

Figure 4.1 shows a determined set of input stimuli applied to the circuit under test (CUT) and the output response of the circuit corresponding to the test stimuli is compared to a known good response or expected response to determine if the circuit is good or faulty. There may be feature extractor circuits which extract important

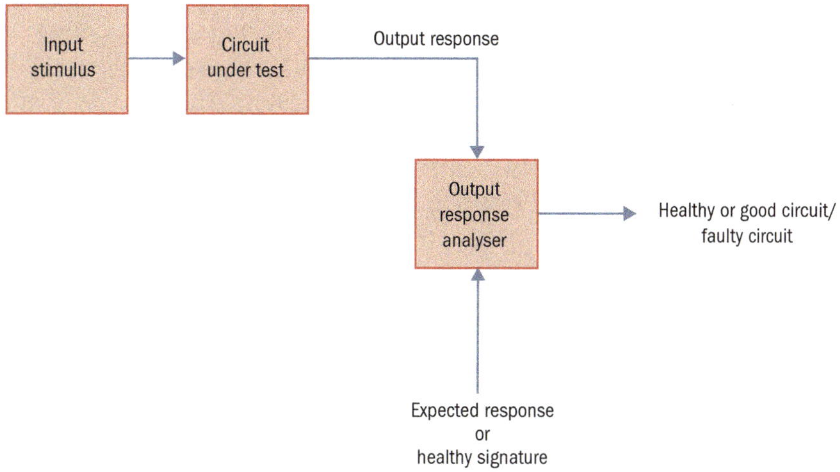

Figure 4.1. BIST basic test flow.

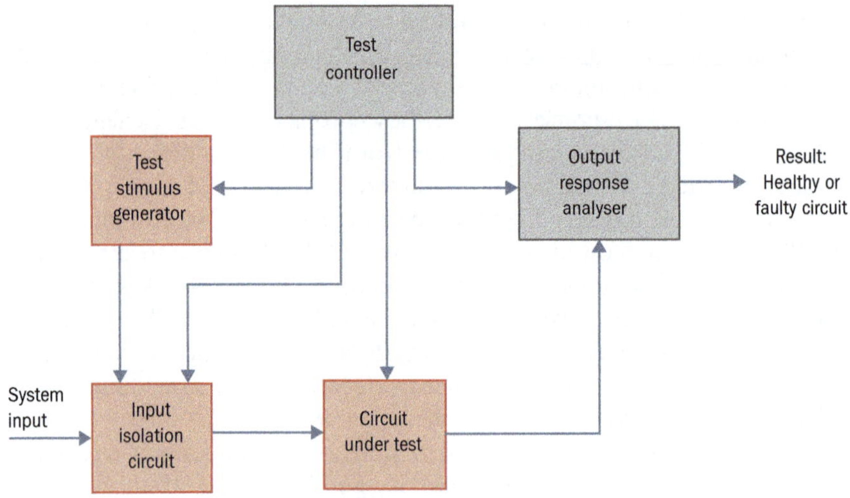

Figure 4.2. General BIST architecture.

information from the output response and then compare them to the expected test parameters.

4.2.3 General BIST architecture

A general BIST architecture is illustrated in figure 4.2.

In figure 4.2 the test pattern generator produces the input stimuli to the CUT and the output response analyser extracts important information and test parameters from the response of the CUT to the test stimuli and generates a 'Pass' or 'Fail' signal after comparison with the nominal or normal or healthy test parameters which are related to a fault-free circuit. The total operation is controlled by the 'test controller' block which is actually the crux of the test system. The entire timing operation and control is carried out by the test controller. The input isolation unit ensures that the system input is not applied to the CUT while the testing is being executed. Only the test pattern generator input should be given to the CUT. This is generally achieved with multiplexers or blocking gates. Apart from input/output pins, the fabrication of a BIST on silicon may also require additional pins for activating the BIST sequence (BIST start), reporting the results of the BIST (pass/fail indication) and an optional pin for a flag that will report the end of BIST operation.

4.3 Test stimulus generation

4.3.1 Principles of on-chip stimulus generation

Generation of an analog stimulus for a BIST is a very important topic in the domain of fault diagnosis and testing. One of the most common methods for the generation of sine wave signals is the closed-loop oscillator [7], which consists of a filtering section with a non-linear feedback mechanism in the arrangement shown in figure 4.3. The quality of the generated signal depends on the linearity and selectivity

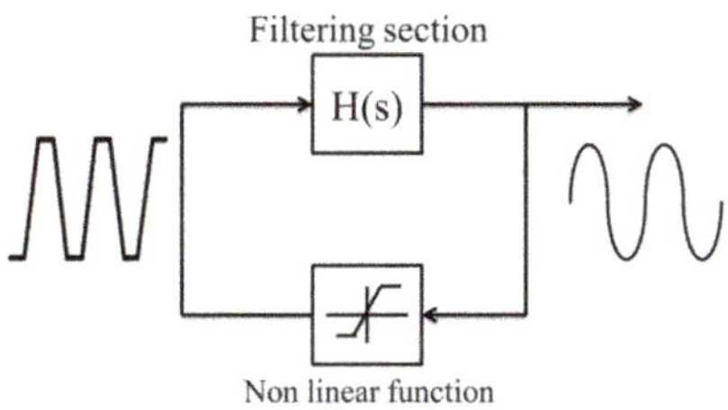

Figure 4.3. Circuit schematic of a signal generator.

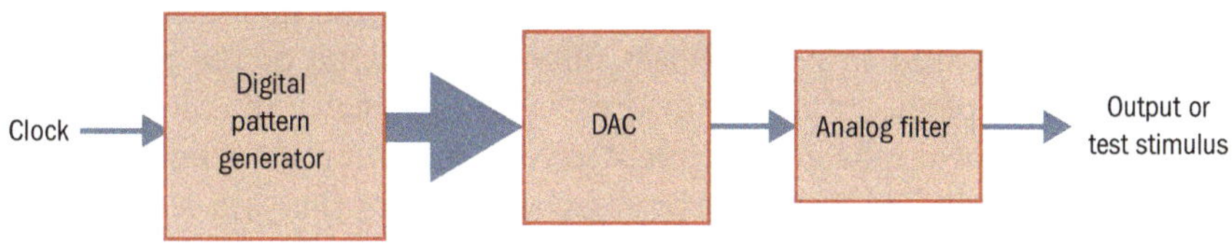

Figure 4.4. Typical open-loop signal generator.

of the filter and the shape of the non-linear function. Highly selective or high Q (quality factor) filters and smooth non-linear functions are needed for the generation of high accuracy, low distortion waveforms [8].

The on-chip generation of sinusoidal test stimulus in the analog domain has a potential application in the field of analog and mixed-signal testing. In fact, most of the analog and mixed-signal subsystems in an SOC, such as analog filters, ADCs, signal conditioners, etc, can be characterized by applying a sinusoidal signal as a test stimulus and analysing the response using a response analyzer. Therefore, the generation of a spectrally pure sinusoidal stimulus with minimum chip area overhead is a challenging task for a VLSI circuit designer.

Most of the proposed strategies for on-chip generation of test stimuli adopt the open-loop scheme [7], as shown in figure 4.4. These waveform generators are typically derived from a digital pattern generator followed by a DAC. The digital pattern generator outputs a digital sequence, the DAC converts the digital pattern to an analog signal, and finally the signal is fed to an analog filter that attenuates all the non-desired components in the output signal [8, 9]. It has the advantages of a digital interface for control and programming tasks, and it also offers the reconfiguration characteristics of digital circuitry.

The proposed approaches achieve the desired signal generation using a variety of different techniques. However, some common elements can be identified:

- Both closed-loop and open-loop methods of generating test signals make use of analog filters to smooth the output signal by eliminating unwanted signal, attenuating all the non-desired frequency components and, in the digital case, reconstructing the analog signal after the DAC interface. This analog output filter is essential for the generation of single-tone analog sine waves, where a given spectral purity has to be assured.

- For an open-loop signal generator, which is usually the preferred solution, the filter should be linear. Otherwise, the linearity of the filter will limit the performance of the generator.

4.3.2 Non-linear wave shaping circuit

It is observed that in a closed-loop oscillator the frequency and amplitude of the output depend on the filter time constant and its gain. Therefore, only a band-pass filter is suitable for this kind of oscillator. On the other hand, in open-loop oscillators a low-pass filter is sufficient with lower order, approximately half of the band-pass filter, because of the absence of zeros in a low-pass filter and hence the better roll-off characteristics compared to a band-pass filter [9, 10]. One of the most important factors for BIST implementation is the area overhead, hence the filter order should be minimum. Moreover, one has to choose among the different technologies, namely active RC, g_m–C, switched capacitor or switched current, that will lead to minimum chip overhead. The area occupied by the digital pattern generator and DAC should also be minimum.

In the conventional oscillator of figure 4.3, the comparator generates the square wave which has odd harmonics with the fundamental frequency. Higher than fifth-order harmonics are easily rejected by the band-pass filter (BPF), hence their contribution to linearity performance is minimal in nature and can be neglected. Thus, the linearity of the oscillator is mainly determined by the third- and fifth-order harmonics, since they are close to the fundamental frequency and have high magnitudes. One researcher devised a new approach where the linearity of the oscillator is improved without requiring a high pole selectivity (Q_p) band-pass filter. This is accomplished by non-linear shaping which consists of a harmonic-suppression technique in a multilevel comparator. The operation of the multilevel comparator completely eliminates the third- and fifth-order harmonics. Therefore, the harmonics in the oscillator's output in figure 4.3 are dependent on the pole selectivity of the band-pass filter and multilevel comparator [8].

4.3.3 Analog filter

An analog signal is a measureable physical quantity whose magnitude varies with time and conveys information about a process or event. The key features of an analog signal are its amplitude, phase and frequency. It will be called a continuous time circuit if the network under consideration processes the signal without sampling it. If the analog signal is sampled at a regular and specific interval a discontinuous signal will be obtained. These signals will then be called discrete time signals and they are not digital signals since their amplitude will vary even in the discrete time domain. The circuits that deal with the discrete time signal are also called discrete time or discrete time domain circuits and even then they are analog circuits. The switched capacitor or switched current circuits are good examples of such circuits [31].

4.3.3.1 Fundamentals of an active filter

An active filter is a circuit used to process the frequency spectrum of an electrical signal. These networks are an essential part of analog signal processing. The biquadratic filter or biquad is basically a second-order filter. It forms the building blocks for higher-order filters. Therefore, emphasis is given to biquads. Active filters employ an operational amplifier (op-amp) as an active device along with resistors and capacitors and dispense of inductors. They are widely used in communications and instrumentation applications.

4.3.3.2 Filter types

The filters are frequency selective networks and they are categorized as low-pass (LP), band-pass (BP), high-pass (HP), low-pass notch (LPN), high-pass notch (HPN) and all-pass (AP). In this section we define each of these filter functions and show their requirements, gain response and pole zero plot.

A biquadratic filter function is expressed by

$$T(s) = K\frac{s^2 + \frac{\omega_z}{Q_z}s + \omega_z^2}{s^2 + \frac{\omega_p}{Q_p}s + \omega_p^2},$$

where

ω_z = zero frequency,
Q_z = zero selectivity,
ω_p = pole frequency,
Q_p = pole selectivity,
K = gain constant.

4.3.3.3 Low-pass filter

A low-pass filter will pass a low frequency from dc to some pre-specified cut-off frequency and will attenuate high frequencies. The requirements for a low-pass filter function are shown in figure 4.5. The low-pass filter is specified by cut-off frequency ω_c, stop-band frequency ω_s, dc gain, pass-band ripple and stop-band attenuation. All other filter functions are specified in the same manner. The low-pass function of figure 4.5 has a pass-band from dc to ω_c, a stop-band from ω_s to infinity and a transition band from ω_c to ω_s. A second-order low-pass filter function is given by the following transfer function

$$T(s) = \frac{d}{s^2 + as + b}. \tag{4.1}$$

Sketches of a second-order low-pass function and its pole zero plot are shown in figures 4.6(a) and (b), respectively. The positions of the poles of equation (4.1) determine the response of the filter in the pass-band. For high $Q_p \left(=\frac{\sqrt{b}}{a}\right)$, a bump occurs in the pass-band at ω_p $(=\sqrt{b})$.

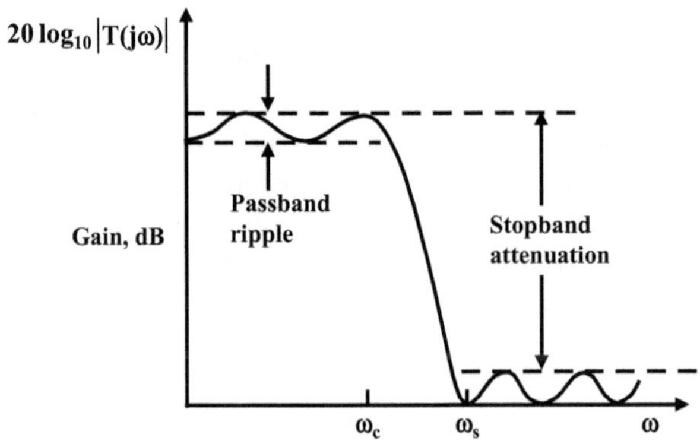

Figure 4.5. Low-pass filter requirement.

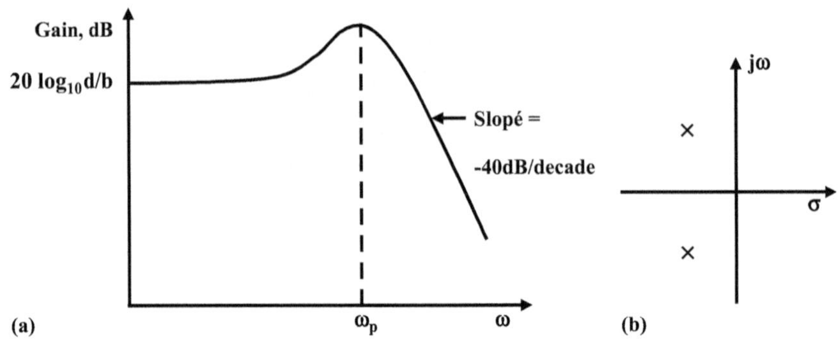

Figure 4.6. (a) Second-order low-pass gain function. (b) Pole zero plot.

4.3.3.4 High-pass filter

A high-pass filter will pass high frequencies above some pre-specified cut-off frequency ω_c and will attenuate low frequencies from dc to some specified stop-band frequency ω_s. The requirement for a high-pass function is shown in figure 4.7. The high-pass filter of figure 4.7 ideally has a pass-band from ω_c to infinity. However, due to the finite gain bandwidth product of the op-amp the pass-band is limited.

A second-order high-pass filter function is given by the following transfer function:

$$T(s) = \frac{ms^2}{s^2 + as + b}. \tag{4.2}$$

Sketches of a second-order high-pass function and the corresponding pole zero plot are shown in figures 4.8(a) and (b), respectively.

4.3.3.5 Band-pass filter

A band-pass filter will pass a band of frequencies while attenuating both lower and higher frequencies. This filter has two stop bands from dc to ω_{s_1} and ω_{s_2} to ∞

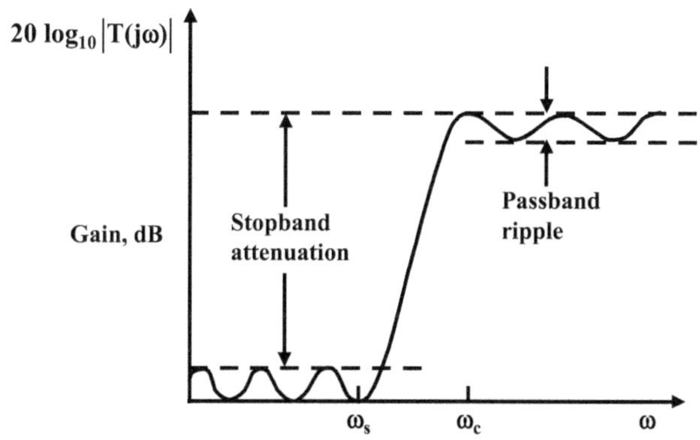

Figure 4.7. High-pass filter requirement.

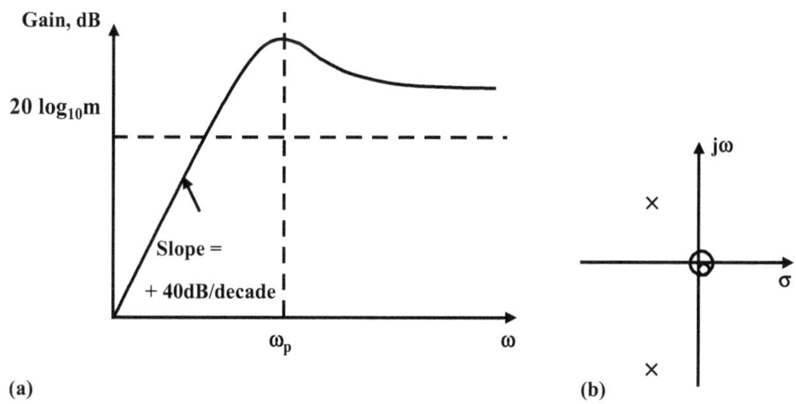

Figure 4.8. (a) Second-order high-pass gain function. (b) Pole zero plot.

(ideally). The pass-band is from ω_1 to ω_2. Obviously there are two transition bands from ω_{s_1} to ω_1 and ω_2 to ω_{s_2}.

The requirements for a band-pass function are shown in figure 4.9. A second-order band-pass filter function is given by the following transfer function:

$$T(s) = \frac{cs}{s^2 + as + b}. \tag{4.3}$$

Sketches of a second-order band-pass function and its pole zero plot are shown in figures 4.10(a) and (b), respectively.

4.3.3.6 Low-pass notch and high-pass notch

The low-pass and high-pass notch filters are special kinds of band-reject filters. A band-reject filter will attenuate a finite band of frequencies while passing both lower and higher frequencies. As a result there are two pass-bands: dc to ω_{c_1} and ω_{c_2} to ∞. The stop-band is from ω_{s_1} to ω_{s_2}. The requirements for a band-reject function are

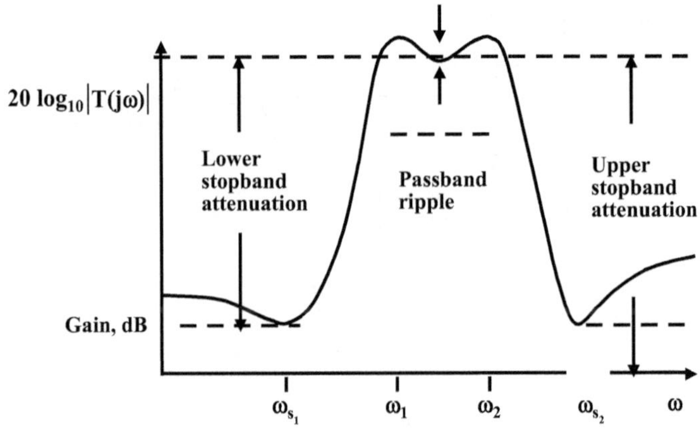

Figure 4.9. Band-pass filter requirement.

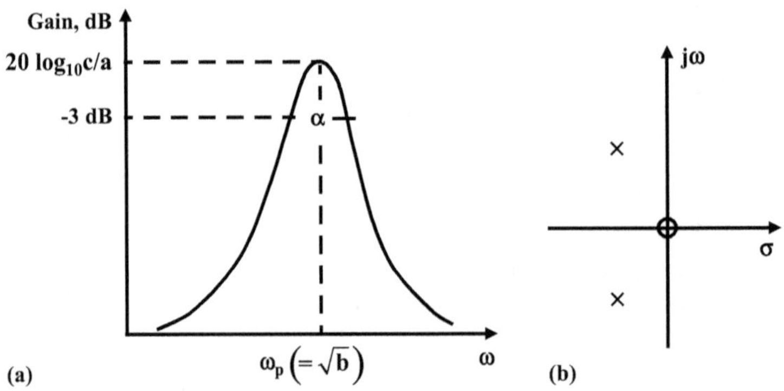

Figure 4.10. (a) Second-order band-pass gain function. (b) Pole zero plot.

shown in figure 4.11. A second-order band-reject filter function has the following transfer function:

$$T(s) = \frac{s^2 + d}{s^2 + as + b}. \tag{4.4}$$

If $d = b$, it is called a band-reject function. If $b < d$ it is a low-pass notch and it is called a high-pass notch when $b > d$. The second-order band-reject gain function and its pole zero plot are shown in figures 4.12(a) and (b), respectively.

The response of second-order low-pass notch and high-pass notch functions and their pole zero plots are shown in figures 4.13(a), (b) and figures 4.14(a), (b) respectively.

4.3.3.7 All-pass filter and delay equalizer

So far we have not considered the phase or delay response of a filter. In the digital transmission of signals, the phase or delay introduced by a filter can cause intolerable distortions of time domain digital signals.

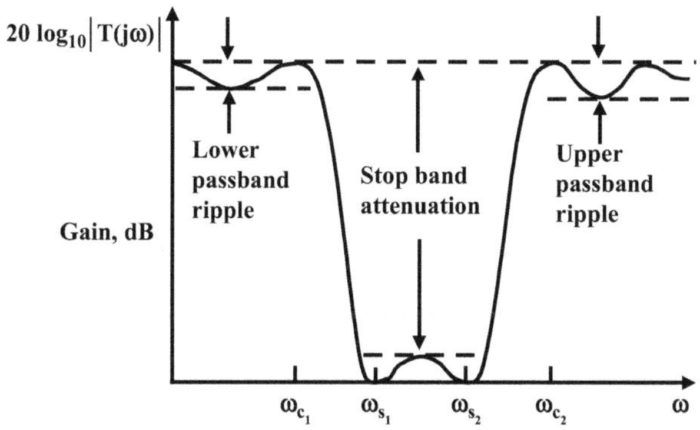

Figure 4.11. The requirements for a band-reject filter function.

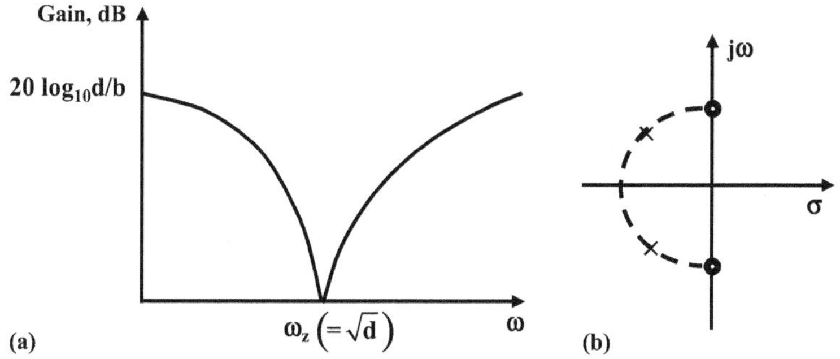

Figure 4.12. (a) Second-order band-reject gain function. (b) Pole zero plot.

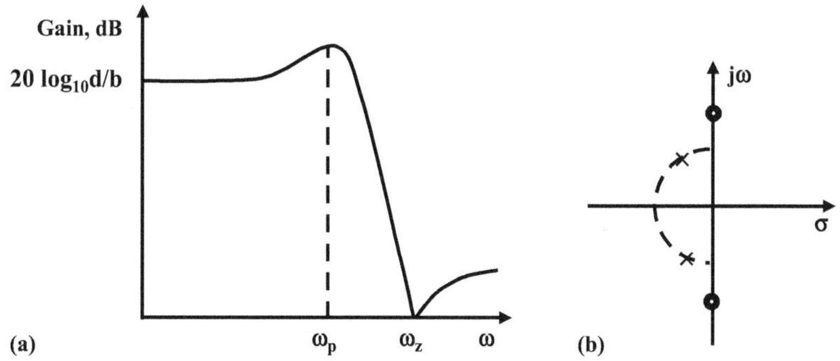

Figure 4.13. (a) Second-order low-pass notch gain function. (b) Pole zero plot.

The delay of a network is defined as

$$\text{delay} = \frac{d}{d\omega}(-\phi(\omega)),$$

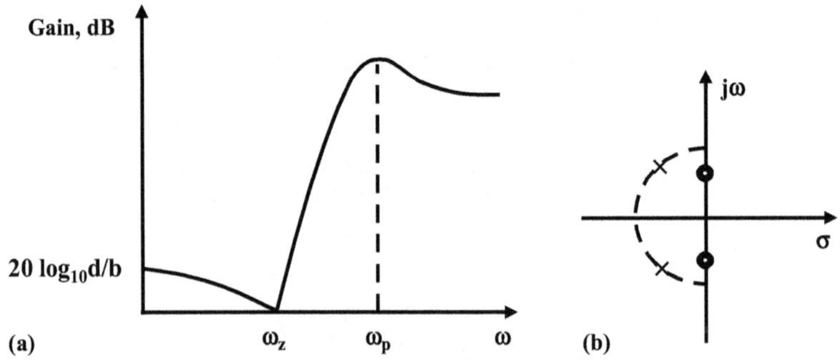

Figure 4.14. (a) Second-order high-pass notch gain function. (b) Pole zero plot.

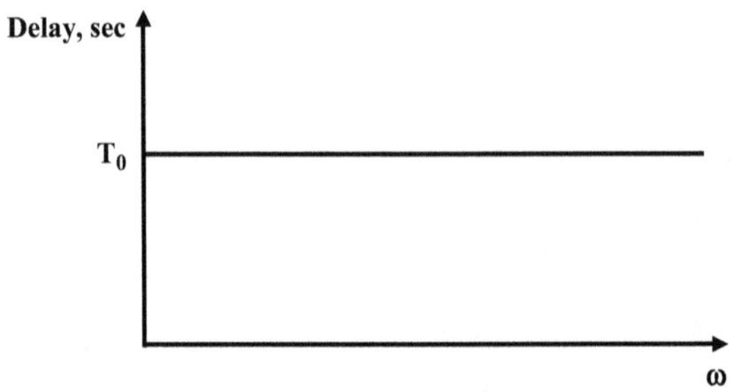

Figure 4.15. The ideal delay characteristic.

where $\phi(\omega)$ is the phase of the gain function. If the transfer function of a filter is expressed in factored form as

$$T_s = \prod_{i=1}^{N} \frac{m_i s^2 + c_i s + a_i}{n_i s^2 + a_i s + b},$$

then the total delay (D) introduced by the network function will be

$$D = \sum_{i=1}^{N} - \frac{c_i(d_i + m_i\omega^2)}{(d_i - m_i\omega^2)^2 + c_i^2\omega^2} + \frac{a_i(b_i + n_i\omega^2)}{(b_i - n_i\omega^2)^2 + a_i^2\omega^2}.$$

The purpose of the delay equalizer or all-pass filter is to introduce a necessary delay at a specified frequency to make the total delay (filter and equalizer) as flat as possible within the frequency range of interest. In this process the delay equalizer must not perturb the gain characteristic of the filter. The ideal delay characteristic of a filter is shown in figure 4.15.

The transfer function of a second-order delay equalizer is as follows:

$$T(s) = K\frac{s^2 - as + b}{s^2 + as + b}.$$

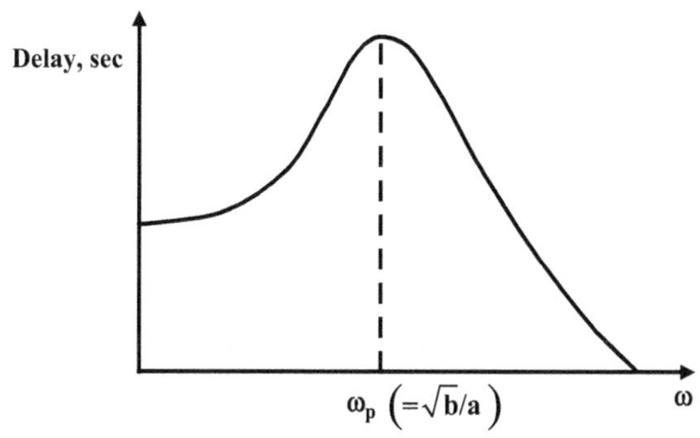

Figure 4.16. Delay characteristic of a second-order delay equalizer.

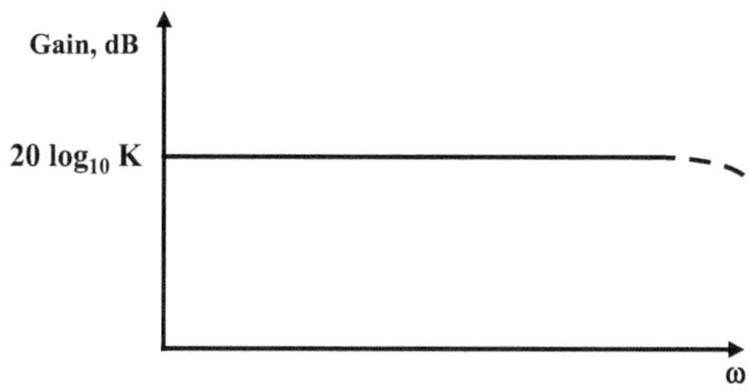

Figure 4.17. Gain characteristic of a second-order delay equalizer.

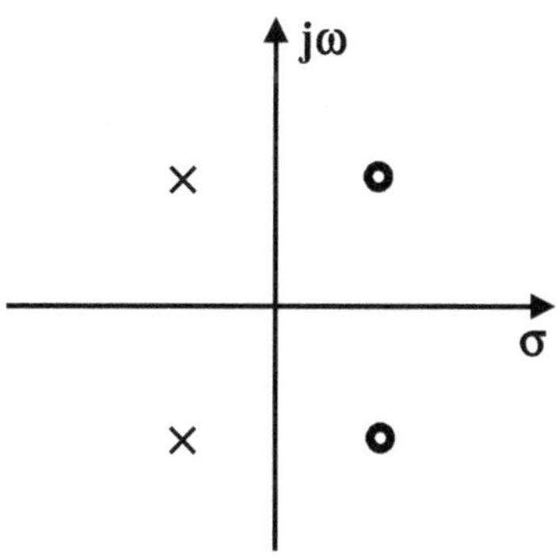

Figure 4.18. Pole zero plot of an all-pass filter.

The delay and gain characteristics of a second-order delay equalizer are shown in figures 4.16 and 4.17, respectively.

Since the gain characteristic in figure 4.17 is flat for the whole frequency range of interest, it is often referred to as an all-pass filter. The pole zero plot of the second-order all-pass function is shown in figure 4.18. The complex conjugate poles and zeros of a second-order all-pass function are symmetrical about the $j\omega$-axis. This is necessary to have a flat gain characteristic.

4.3.3.8 State variable structure

A second-order analog filter can be realized in a single-amplifier, two-amplifier or three-amplifier topology. However, a three-amplifier structure has the advantage that it has the capability of tuning the filter parameters independently. A three-amplifier state variable filter is shown in figure 4.19.

Analysis of the circuit of figure 4.19 yields

$$\frac{V_0}{V_{\mathrm{IN}}}(s) = -\frac{R_8}{R_6}\frac{s^2 + s\left(\frac{1}{R_1C_1} - \frac{R_6}{R_4R_7C_1}\right) + \frac{R_6}{R_3R_5R_7C_1C_2}}{s^2 + \frac{1}{R_1C_1}s + \frac{R_8}{R_2R_3R_7C_1C_2}}. \tag{4.5}$$

From equation (4.5) the expressions for ω_p, Q_p and K can be written as

$$\omega_p = \sqrt{\frac{R_8}{R_2R_3R_7C_1C_2}}, \quad Q_p = R_1\sqrt{\frac{R_8C_1}{R_2R_3R_7C_2}}$$

and

$$K = -\frac{R_8}{R_6}.$$

Comparing with the general biquadratic the design equations are

$$C_1 = C_2 = 1,\ R_2 = R_3 = R_7 = R_8 = R = \frac{1}{\sqrt{b}},$$

$$R_1 = 1/a,\ R_4 = 1/K(a - c),\ R_5 = \frac{\sqrt{b}}{Kd},\ R_6 = \frac{1}{K\sqrt{b}}.$$

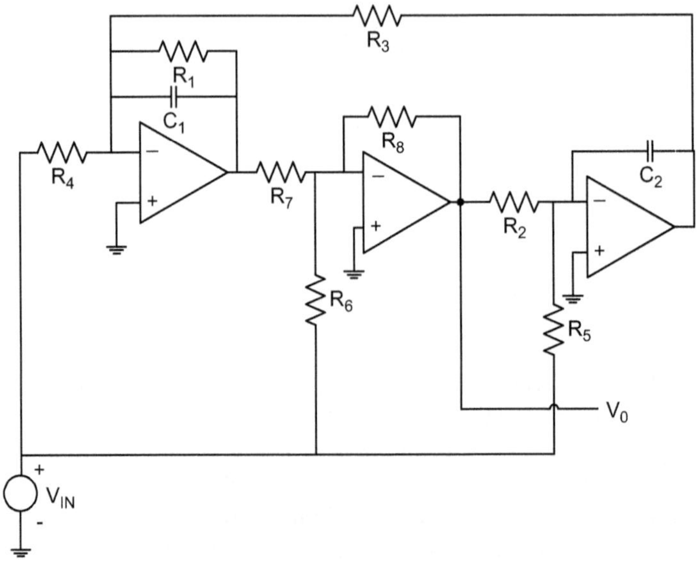

Figure 4.19. Three-amplifier feedforward second-order filter.

Here also the design equations yield positive element values for $a \geqslant c$.

Despite the advantages of having large dynamic ranges and precision output, state variable filters suffer from disadvantages in terms of the requirements for large component spread and a large area of silicon due to the on-chip fabrication of resistances. If the filter is designed for lower cut-off frequencies of a few hundred hertz, it requires a resistance of the order of a few mega-ohms and a capacitance of a few nano-farads. In CMOS technology, obtaining such a high component spread requires a huge area, so an active RC filter is not suitable for integrating on-chip applications, hence the filter topology has to be changed.

An alternative solution is to perform the filtering operation with a switched capacitor filter, which will be discussed in the next subsection of this chapter.

4.3.4 Switched capacitor filter

A switched capacitor is treated as an approximate equivalent to a resistor, thereby opening up the whole active RC design literature for these devices. Careful observation shows that a switched capacitor circuit is, in fact, a sampled data network and as such it is more accurately described in terms of the z transform variable instead of the s transform. The familiar bilinear z transformation can be applied either to an appropriate analog transfer function or to an analog circuit to begin switched capacitor design. In the synthesis process the circuit parameters can be precisely achieved by controlling the capacitor ratios only. Finally, the physical size of all the components can be made sufficiently small, so that a high quality circuit can be laid out on a single monolithic semiconductor chip. The basic motivation for the development of the switched capacitor filter was the need to obtain a fully integrated frequency selective device on a single chip. Integrated circuits cannot contain high Q inductors, since the quality factor Q decreases with decreasing size. Resistors can easily be integrated, but occupy a large area on the chip and their tolerances and temperature coefficients do not track with those of the capacitor on the same chip. Hence the filter parameters cannot be accurately controlled in RC circuits. High quality capacitors, in contrast, can be realized conveniently in an MOS integrated circuit. The dielectric material is SiO_2, an excellent insulator. The electrodes or plates of the capacitor can be made of metal or polycrystalline silicon, or heavily doped crystalline silicon.

The good switches can also be fabricated using MOS technology. The on-resistance depends on the area allowed for the MOS transistor used as the switch. The off-resistance is, for all practical purpose, infinite. The stray capacitance between the gate and the drain and source are of the order of 2–10 fF. It can play an important role in causing clock signal feed-through and dc offset. A key component in the switched capacitor circuit is the active element which is invariably an operational amplifier (op-amp). In terms of chip area, an op-amp occupies about as much as a 50 pF capacitor. We can simulate inductors or frequency-dependent negative resistors (FDNRs) using capacitors, switches and the op-amp. The gain of the circuit or the position of poles and zeros will depend only on the ratios of capacitances which can be precisely controlled.

4.3.4.1 Switched capacitor resistor

As shown in figure 4.20 a capacitor is connected between two nodes through a switch that can be implemented by an MOS transistor. The voltages v_1 and v_2 are connected to nodes 1 and 2, respectively. It can easily be shown that the average current i_2 delivered by the capacitor to node 2 is $(C/T_c)(V_1 - V_2)$, where $f_c = 1/T_c$ is the clock frequency that is applied to the control terminal of the switch and T_c is the clock period. Therefore, the circuit of figure 4.20 is equivalent to a resistor of value T_c/C connected between nodes 1 and 2. However, the equivalence between the resistor and the circuit of figure 4.20 is valid only for signal frequencies satisfying $f \ll f_c$.

4.3.4.2 Resistor simulation

In figure 4.20 the switches will be closed alternately by the non-overlapping clocks ϕ_1 and ϕ_2. We assume the switches are closed (or open) for T_1 and T_2 seconds. The time period of the clock is $T_c = T_1 + T_2$ seconds. For a 50% duty cycle clock $T_1 = T_2 = \frac{T_c}{2}$ seconds. The voltage sources $V_1(t)$ and $V_2(t)$ are assumed to be ideal and $V_1 > V_2$. With ϕ_1 ON, the capacitor is charged to V_1. At $t = T_1 = \frac{T_c}{2}$, ϕ_1 is OFF and ϕ_2 is ON and the capacitor C discharges to V_2 volts. The cycle is repeated at $t = T_c = T_1 + T_2$ seconds. For an complete cycle the charge Q transferred from node 1 to node 2 is C $(V_1 - V_2)$. Therefore the net current flows as

$$I_{\text{net}} = \frac{Q}{T_c} = \frac{C}{T_c}(V_1 - V_2). \tag{4.6}$$

Therefore,

$$I_{\text{net}} = C \cdot f_c(V_1 - V_2), \tag{4.7}$$

where f_c is the frequency of the clock.

The current I_{net} passes through $R_{\text{equivalent}}$, where

$$\frac{V_1 - V_2}{R_{\text{equivalent}}} = C \cdot f_c(V_1 - V_2), \tag{4.8}$$

or

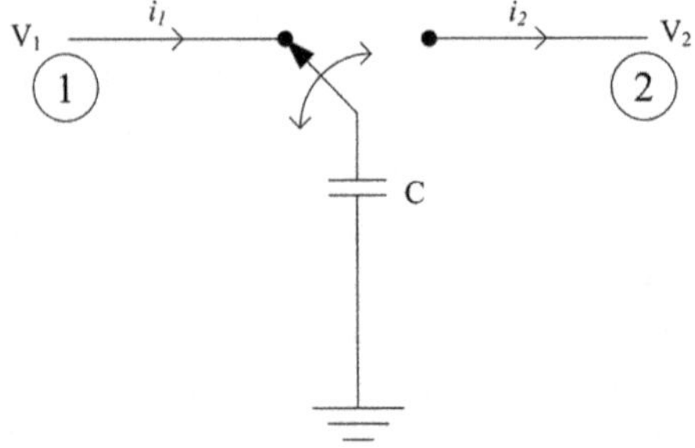

Figure 4.20. Switched capacitor equivalent of a resistor.

$$R_{\text{equivalent}} = \frac{1}{C_{f_c}}. \tag{4.9}$$

From the above considerations we can arrive at the conclusion that by charging and discharging a capacitor continuously a resistor can be simulated. One of the primary requirements for the circuit to work properly is that the clock frequency needs to be much higher than the signal frequency, as demonstrated above.

Two non-overlapping clocks are required for the switched capacitor circuit operation. Non overlapping clocks are shown in figure 4.21. The frequency of the clock should be at least ten times greater than the frequency of the signal.

4.3.4.3 Switched capacitor integrator

The method of designing a switched capacitor filter is based on replacing the integrators in the state variable and the leap-frog active filters by switched capacitor integrators. The resulting second-order filter is composed of two integrators—one is an inverting and the other is a non-inverting integrator. As an inherent property of the switched capacitor circuit, its transfer function is completely independent of the absolute value of the passive components, instead it depends solely on the capacitor ratios which can be precisely achieved in today's MOS technology. In figures 4.22(a) and (b) a non-inverting and an inverting integrator are shown, respectively. Their transfer functions are completely independent of the stray capacitances between any node and ground. This helps the circuit designer use a very low value of capacitance, thus reducing the chip area on silicon. The integrators shown in figures 4.22(a) and (b) are not perfect integrators since their transfer functions are not in the form $1/j\omega\tau$. The two clock pulses φ_1 and φ_2 are 90° out of phase. The non-overlapping 50% duty cycle clock pulses are shown in figure 4.21. However, the gain of the integrator transfer function is not proportional to $1/\omega$.

The transfer function of the inverting integrator is

$$\frac{V_0}{V_{\text{in}}}(z) = -\frac{k}{1 - Z^{-1}} \tag{4.10}$$

and that of the non-inverting integrator is

$$\frac{V_0}{V_{\text{in}}}(z) = \frac{kZ^{-1}}{1 - Z^{-1}}. \tag{4.11}$$

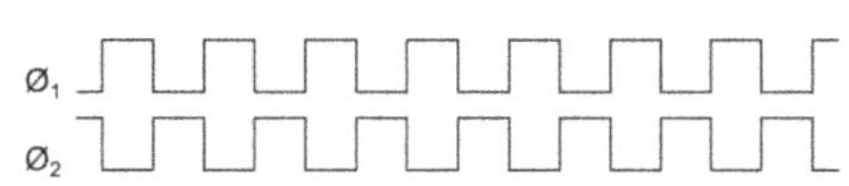

Figure 4.21. Non-overlapping clock.

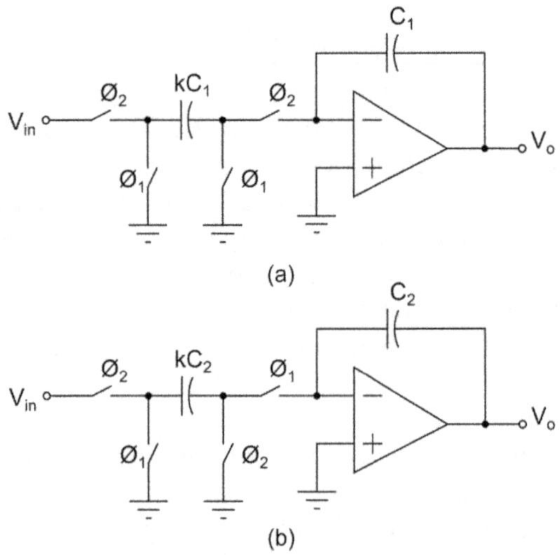

Figure 4.22. (a) Inverting integrator and (b) non-inverting integrator. Since the integrators are sampled data circuits, the transfer function is to be written in the *z*-domain instead of the *s*-domain.

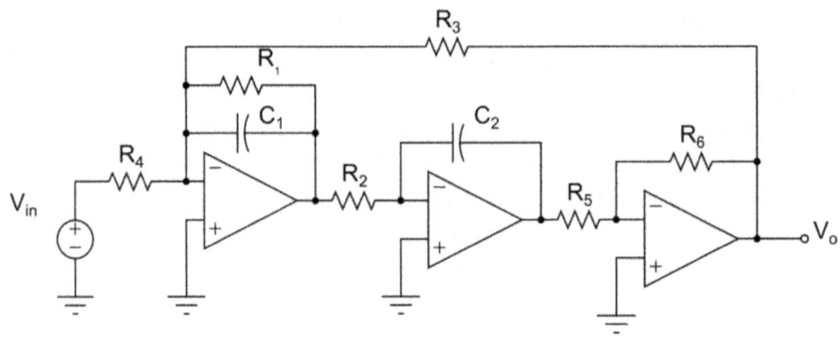

Figure 4.23. An RC state variable filter that will realize the band-pass and low-pass function.

4.3.4.4 Switched capacitor state variable second-order filter circuit

The biquad circuit is based on two integrator loops. To obtain a finite pole selectivity one of the integrators must be leaky or damped. A second-order filter circuit realized with resistance and capacitance is shown in figure 4.23.

At the output of the first amplifier we shall obtain a band-pass function. A low-pass and inverted low-pass function will be available, respectively, at the output of the second and third amplifiers. An inverting function is necessary to avoid 360° or 0° phase shift before overall feedback is given through the resistor R_3. The switched capacitor realization of the second-order filter circuit is shown in figure 4.24. The function of the integrator and inverter of the RC circuit will be served by one non-inverting switched capacitor integrator, making a considerable saving of area on silicon. Moreover, the circuit of figure 4.24 is switch minimized. Careful observation of the circuit of figure 4.24 shows that each of the resistors R_1 and R_3 of figure 4.23

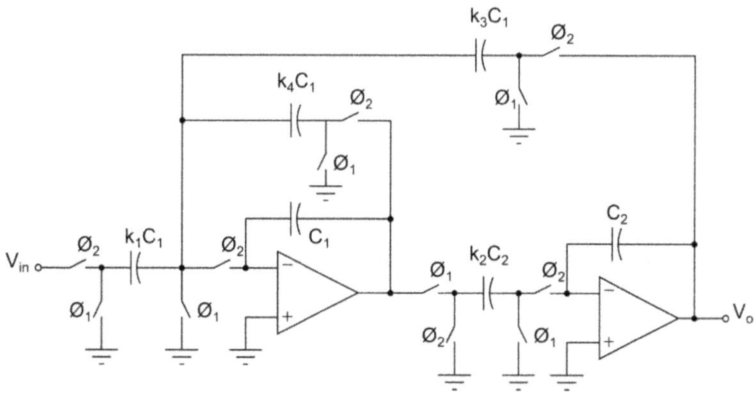

Figure 4.24. Switched capacitor realization of the circuit of figure 4.23.

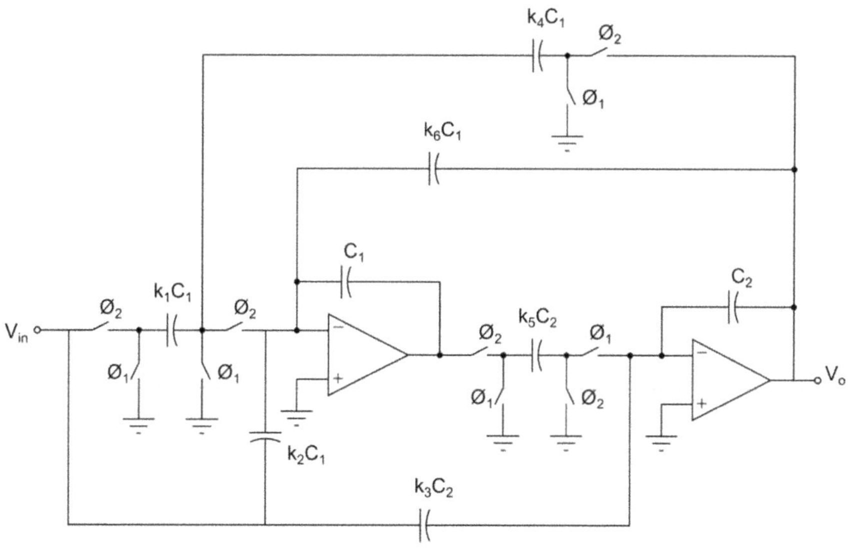

Figure 4.25. A switched capacitor-based second-order filter which can realize any filter function.

are replaced by two switches and one capacitor instead of four switches and one capacitor.

A second-order filter circuit that can realize any filter function is shown in figure 4.25. One has to write the difference equation of the individual element of the discrete domain circuit if one wants to find the overall transfer function. It is even easy to find the transfer function of the switched capacitor filter circuit since it comprises integrators in different forms whose z-domain equations have already been established. The signal flow graph of the total switched capacitor-based filter circuit is shown in figure 4.26. The signal flow graph will lead to a transfer function of the circuit as follows:

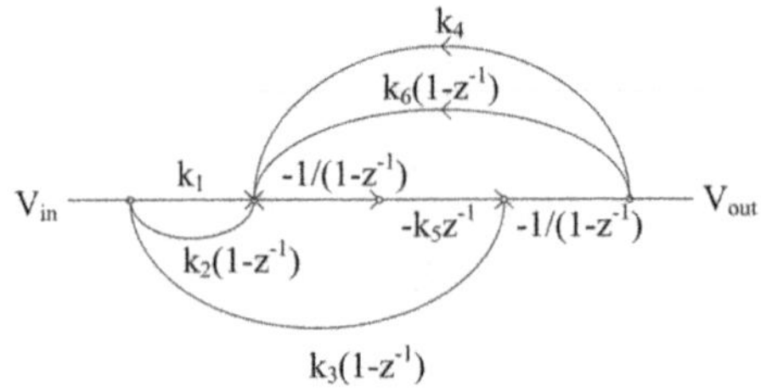

Figure 4.26. Signal flow graph of the switched capacitor filter shown in figure 4.25.

$$\frac{V_0}{V_{in}}(z) = -\frac{k_3z^2 + (-2k_3 + k_1k_5 + k_2k_5)z + (-k_2k_5 + k_3)}{z^2 + (-2 + k_4k_5 + k_5k_6)z + (1 - k_5k_6)}. \tag{4.12}$$

A biquadratic function that can be realized for any filter function in the continuous time domain can be written as follows:

$$H(s) = -\frac{ms^2 + cs + d}{s^2 + as + b}.$$

In the discrete time domain the above transfer function will have the following form:

$$H(z) = -\frac{n(d + c + m)z^2 + 2n(d - m)z + n(d - c + m)}{z^2 + 2n(b - 1)z + n(1 + b - a))}, \tag{4.13}$$

where

$$n = \frac{1}{1 + b + a}.$$

For any switched capacitor circuit the design equations are essentially finding the capacitor ratios. In another way, the positions of the poles and zeroes can be derived by equating the coefficients of the like power of the z of the denominator and numerator in equations (4.12) and (4.13). For poles,

$$k_4k_5 = 4nb \tag{4.14}$$

$$k_5k_6 = 4na. \tag{4.15}$$

As is obvious, it is not possible to solve the equations since there are three unknowns in two equations. To maximize the dynamic range the time constants of two integrators are chosen to be equal.

Therefore,

$$k_4 = k_5.$$

The zeros will determine the filter functions:

- It will be a low-pass filter for $m = c = 0$; then, $k_2 = 0$, $k_1k_5 = 4nd$ and $k_3 = nd$.
- It will be a band-pass filter for $m = d = 0$; then, $k_1 = 0$, $k_2k_5 = 2nc$ and $k_3 = nc$.
- It will be a high-pass filter for $c = d = 0$; then $k_1 = k_2 = 0$ and $k_3 = nm$.

- It will be a band-reject filter for $c = 0$; then $k_2 = 0$, $k_1k_5 = 4nd$ and $k_3 = n(m + d)$.

It is observed that the circuit of figure 4.25 can realize any filter function. Therefore, it is a basic building block for designing any filter function either in cascade, coupled biquad structure or circuit simulation of an LC ladder filter.

On careful observation of figures 4.24 and 4.25 it is revealed that unlike an RC active filter, a switched capacitor in state variable form does not need the third op-amp since a non-inverting integrator can be made simply by changing the clock sequence and this provides considerable savings on silicon area. A fully differential SC filter biquad is also proposed by some researchers [8]. A fully differential SC state variable filter is shown in figure 4.27. The circuit is switch minimized and the integrators and therefore the filter are insensitive to parasitic capacitances. Its performance and noise rejection are somewhat better than the SC state variable filter where op-amps are in single-ended mode.

However, an SC circuit has several limitations for which researchers have proposed alternative solutions.

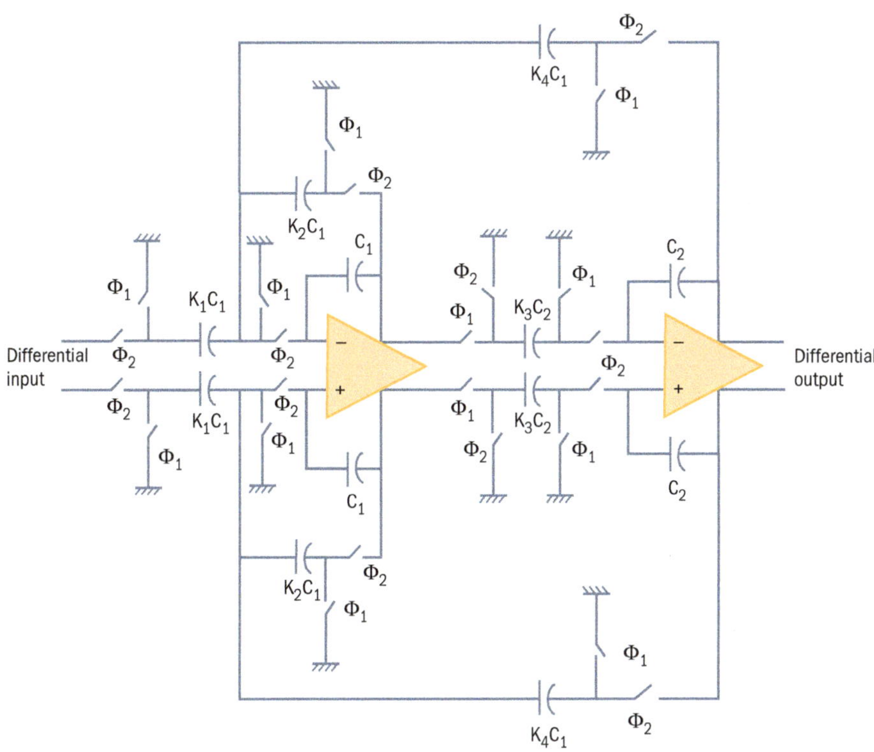

Figure 4.27. A fully differential parasitic insensitive switched capacitor state variable biquad filter with switch minimization.

4.4 OTA-C filter

4.4.1 Operational transconductance amplifier and capacitor filter

The OTA-C filter does not use any resistor or switches; rather it uses an operational transconductance amplifier and capacitor. It utilizes OTA and capacitors to implement integrators. Moreover, it uses the feedback to improve the frequency response. The great advantage of this filter is that filter parameters can be tuned by controlling the current which will vary the transconductance gain (g_m). Sometimes the researchers call it the g_m–C filter. The performance of the amplifier will largely affect the analog filters.

The dynamic range of the OTA-C filter depends primarily on the dynamic range of the OTA. The limited dynamic range of the OTA is confined by the linear input range and noise level, which restricts the dynamic range of the filter, i.e. the performance of OTA-C filters, depends on: (i) the OTA itself, which is the main source of noise and distortion in the filter and (ii) the OTA-C filter topology. The fully differential transconductor–capacitor filter is shown in figure 4.28, where all transconductors have equal gain (g_m). It has differential input and differential output. This will make the design easier. The distortion introduced by the OTA depends upon the linearity of that OTA; hence good design of the OTA ensures the better performance. In practice, mismatch in transconductance parameters will occur due to faults during fabrication, which may cause performance change. Thus the sensitivity of the transfer function to the fault error must be small [11–13].

4.4.2 Designing the OTA

The high performance voltage-to-current converter circuit which can achieve a wide transconductance tuning range is designed. The working mode of the transconductor can be set in different inversion regions and the transconductance can be tuned

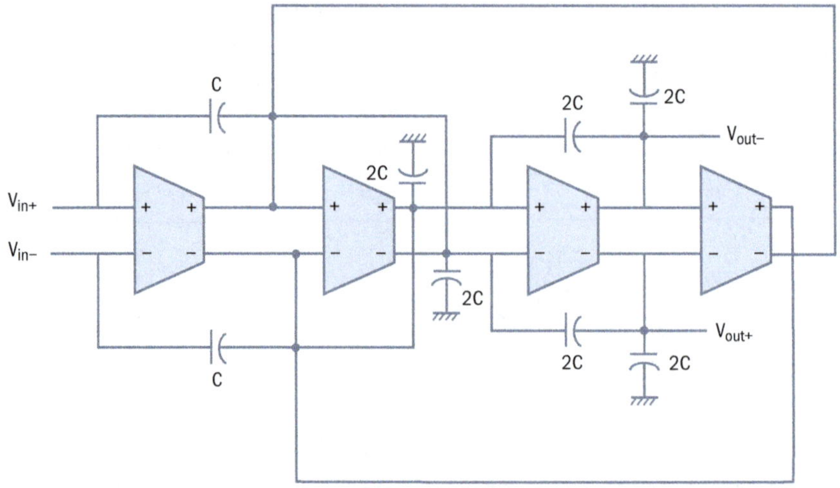

Figure 4.28. Fully differential OTA-C second-order Butterworth low-pass filter.

widely by changing the dc bias current. An OTA is used for the design of the second-order low-pass filter.

Basically an OTA is a differential-input voltage-controlled current source (DVCCS), and its operation is defined by the following equation:

$$I = g_m(V_1 - V_2), \text{ A.} \tag{4.16}$$

The transconductance g_m can be controlled externally by the bias current. These OTAs have certain advantages over op-amps as well as disadvantages. The main advantage of the OTA is their wider bandwidth, which makes them more useful than the op-amps in the design of active filters operating at high frequencies (up to the gigahertz region), since an op-amp's high frequency performance is limited by its gain bandwidth product. Moreover, it has the advantage of on-chip tuning by varying the bias current of the OTA. However, they have some drawbacks, one of which is the limited range of input voltage for linear operation [13]. Other major imperfections include the finite input and output impedances of the OTA, as well as the frequency dependence of transconductance g_m. Despite all these imperfections, OTAs are very useful for the design of tunable active filters at high frequencies.

The transconductance of the OTA can be adjusted by using the bias current through the input differential stage. These types of filters can be operated in open-loop conditions and have very high operating ranges (0.1 Hz to 10 GHz) [13] with very low area overhead, but have a very low range of linearity and parasitic effects are also of much concern. Parasitic effects can be nullified by using on-chip tuning, and the linearity can be enhanced by using source degeneration along with current division.

The transconductance amplifier circuit in the differential mode is shown in figure 4.29. It consists of a source-coupled pair with poly-silicon resistors and a cross-coupled high impedance load. The linearity is enhanced by the degenerated resistors. Transistors M_3, M_5, M_6 and M_4 are matched. M_1 and M_2 operate in the triode region, acting as degeneration resistors to provide different operating points at nodes 1 and 2. M_3 and M_4 act like a pair of positive resistors $R+$, while M_5 and M_6 function as negative resistors $R-$. The output impedance of the transconductor depends on the parallel combination of $R+$ and $R-$, the values of which are controlled by voltages V_1 and V_2, respectively. As a consequence, the output impedance and, therefore, the quality factor of the integrator can be made maximum with proper combinations of voltages V_1 and V_2. A tunable integrator for very high frequency integrated filters can be realized by adjusting the gate voltage M_2, which controls the tail current, and thus g_m. Moreover, it will have high speed because of the absence of internal high impedance nodes. It pushes non-dominant poles to the gigahertz range [11]. The gain and phase plots of the OTA are shown in figure 4.30.

The linearity was enhanced by the degenerated poly-silicon resistors, and hence the input common-mode voltage range is also enhanced. The input common-mode range plot for the OTA is shown in figure 4.31. The frequency response and transient response of the OTA-C filter are shown in figures 4.32 and 4.33. The corner frequency of the filter can be varied by varying the transconductance value of the

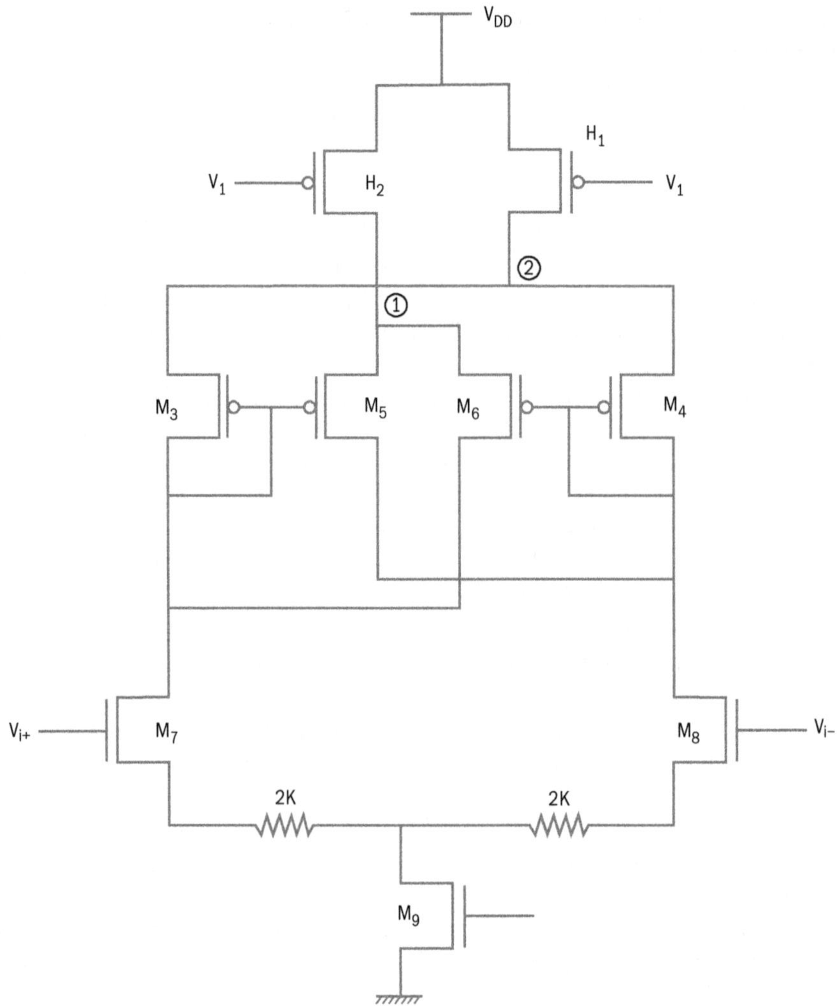

Figure 4.29. Operational transconductance amplifier.

OTA, which can be achieved by varying the bias current, which is dependent on the bias voltage v_{bn}. To change the bias current and hence the transconductance value, there should be a mechanism which can continuously monitor the output of the OTA-C filter. This technique is called auto-tuning.

4.5 Dynamic parameter evaluation of a pipelined ADC

4.5.1 Pipelined ADC

As there has been a lot of advancement in the field of digital signal processing, the ADC has played an important role by acting as an interface between the analog and digital signals. A pipelined ADC is inherently a multi-step amplitude quantizer in which the digitization is performed by a cascade of many topologically similar or

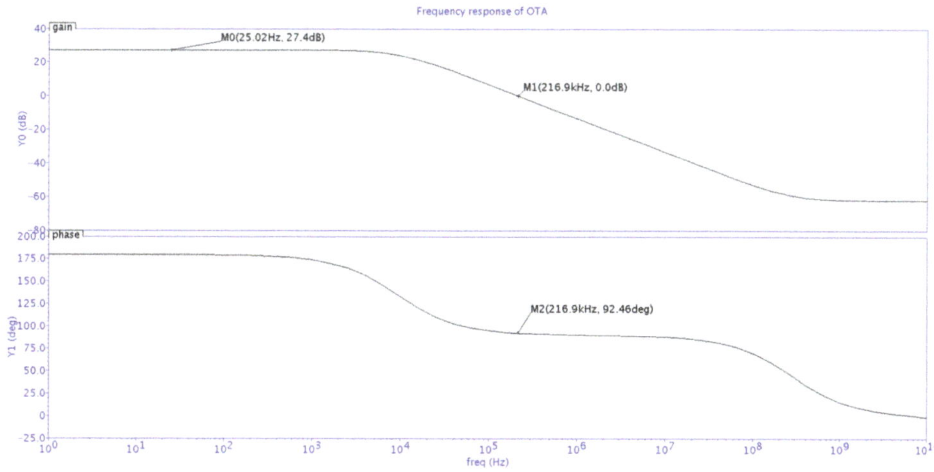

Figure 4.30. Frequency response of the OTA.

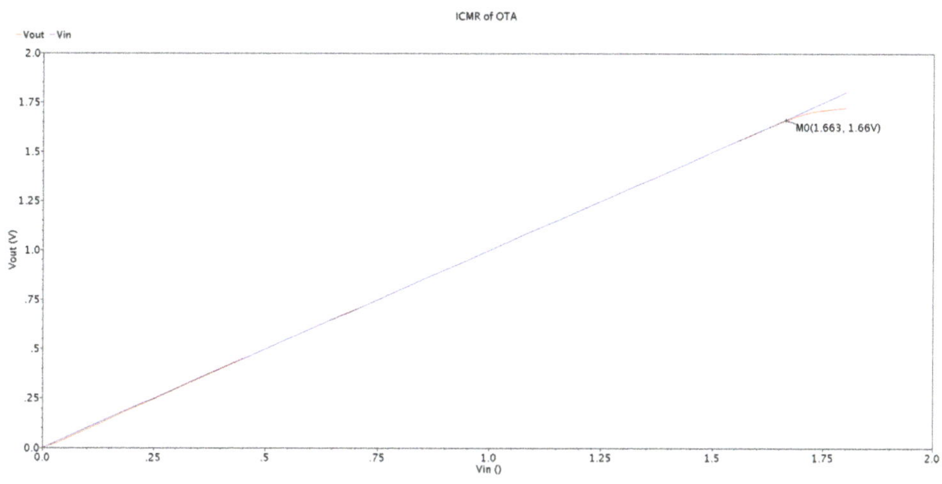

Figure 4.31. The input common-mode range shows the linearity of the OTA.

identical stages of low-resolution analog-to-digital encoders [24, 25]. Pipelining enables high conversion throughput by inserting analog registers, i.e. sample-and-hold amplifiers (SHAs), between stages that allow a concurrent operation of all stages. This is performed at the cost of increased latency. A block diagram of a pipelined ADC is shown in figure 4.34. A pipelined stage takes two actions when an input signal arrives: first the input signal is sampled by the SHA and second a coarse quantization is performed by the sub-ADC. These two operations are often performed simultaneously. The resolution of the conversion is enhanced by passing a residue signal (the unconverted part of the input signal) to the later stages where it is further quantized. A pipelined ADC consists of a coarse comparator and a multiplier DAC (MDAC), which integrates the sample-and-hold (S&H), the DAC,

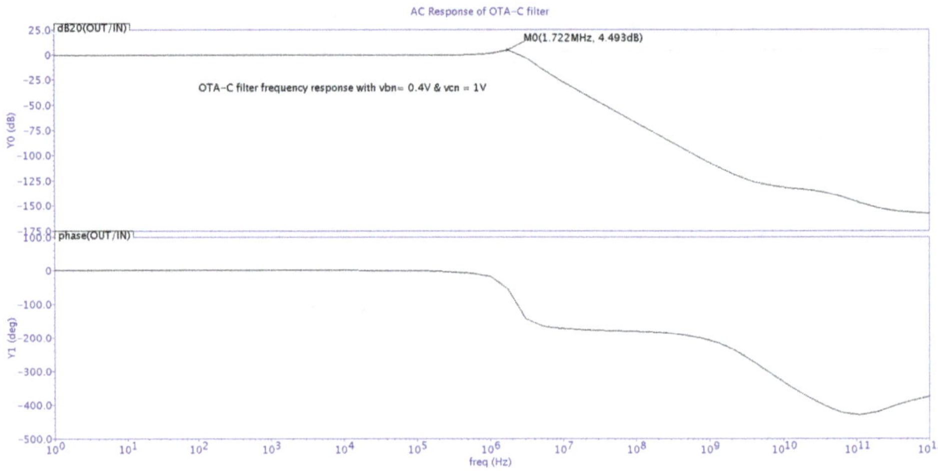

Figure 4.32. Frequency response of the OTA-C filter.

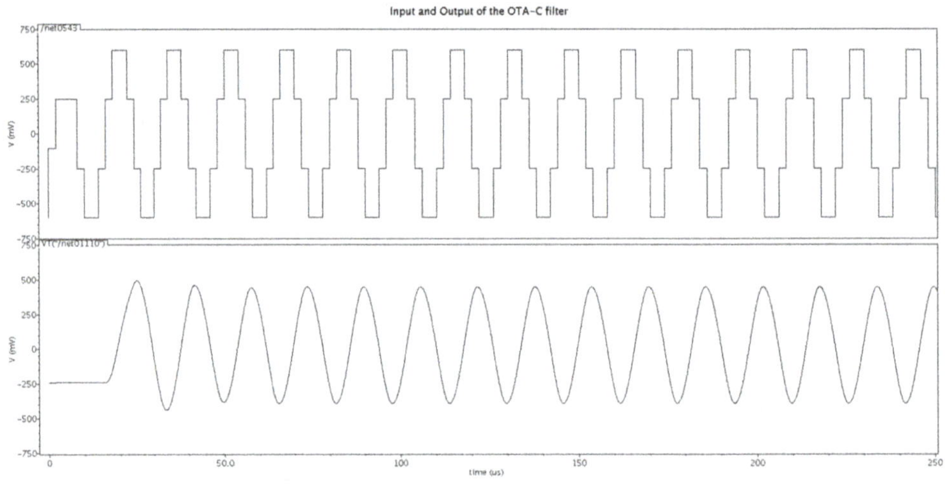

Figure 4.33. Transient response of the OTA-C filter.

the subtraction and the residue-gain functions. This keeps the signal level constant and allows the sharing of an identical reference throughout the pipelined stages. The conversion accuracy thus depends on the precision of the residue signals.

The advantage of this architecture is its reduced complexity and smaller component count at the expense of the speed of conversion. It is less hardware intensive than flash ADC. An ADC of a given resolution can be achieved by cascading an appropriate number of identical pipelined stages. The major disadvantage of this architecture is the latency in the converter. A digital error correction logic is necessary, as shown in figure 4.34 [28].

Pipelined ADCs provide an optimum balance of size, speed, resolution, power dissipation and analog design effort, thus they have become increasingly attractive

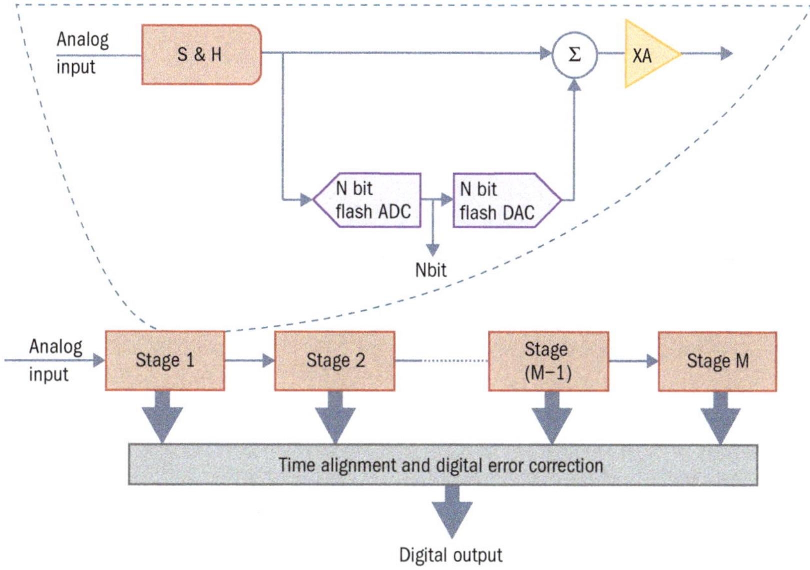

Figure 4.34. Block diagram of the pipelined ADC.

to major data-converter manufacturers and their designers. The pipelined ADC is the architecture of choice for sampling rates up to 100 MSPS. The design complexity increases only linearly (not exponentially) with the number of bits, thus providing converters with high speed, high resolution and low power at the same time. Pipelined ADCs are very useful for a wide range of applications, most notably in digital communications where a converter's dynamic performance is often more important than traditional dc specifications such as DNL and INL [22].

4.5.2 Dynamic performance parameter evaluation

An 8 bit pipelined ADC has been designed in the CMOS 180 nm technology process, as shown in figure 4.34, and has been utilised as the CUT part in the BIST system. The signal that is obtained from the OTA-C filter is fed to the pipelined ADC, which generates an 8 bit digital code for corresponding input. The code obtained from the ADC is passed through an algorithm to reconstruct the digital code [4], and a 1024 point FFT has been calculated from the reconstructed signal, shown in figure 4.35 (only at 100 KHz frequency), from which the dynamic performance parameters have been evaluated. An FFT plot of the reconstructed signal is shown in figure 4.35.

4.6 Fault analysis using the BIST system

4.6.1 Sources of error in a pipelined ADC

ADC non-linearity errors and circuit noise are increased due to degradation of the ideal functionality of the SHA, flash ADC and MDAC circuits and therefore the

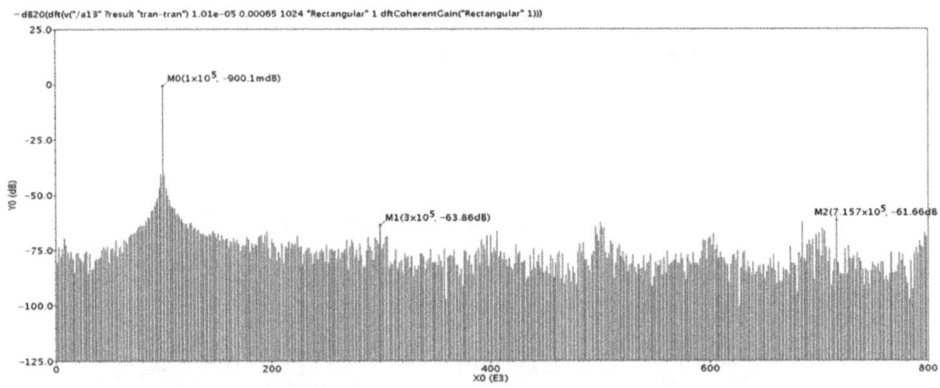

Figure 4.35. An FFT plot of the reconstructed wave of the output of ADC.

final resolution and conversion rate of the ADC are limited. Major non-ideal factors include the switching charge injection, thermal noise, op-amp finite open-loop gain, input parasitic capacitance and output voltage settling error.

4.6.2 Distortion by switching charge injection

As the transmission gates in the SHA and MDAC circuits are switched 'OFF', charges are injected from the transistors' parasitic capacitances onto the sampling capacitors. These injected charges create an error voltage to the voltage across the capacitors. The charge injected (Q) by each of the PMOS and NMOS transistors of the transmission gate is given by

$$Q = WLC_{ox}(V_{GS} - V_{th}), \tag{4.17}$$

where W, L, C_{ox}, V_{GS} and V_{th} are, respectively, the transistor channel width, channel length, gate oxide capacitance, gate to source voltage and threshold voltage of the MOS transistor.

The CMOS transistors of transmission gates have minimum channel length and width to reduce the injected charges and clock feed-through. Furthermore, the entire pipelined ADC is designed to be fully differential. This is because the voltage error due to charge injections of the switches is minimum in fully differential circuits. This is because the added voltage error is a common-mode noise and will be eliminated at the output due to the nature of differential circuits.

4.6.3 Distortion due to common thermal noise

Thermal noise determines the MDAC and SHA minimum sampling capacitor values. As the op-amp capacitance load value decreases, the op-amp slew rate and bandwidth increase, and so does the ADC maximum conversion rate. On the other hand, by decreasing the capacitance load value, the voltage thermal noise on the capacitors increases. If this noise exceeds or is approximately equal to the quantization noise, then the SNR and ENOB will decrease noticeably. It is required to select values for the sampling capacitors that will not noticeably degrade the SNR and ENOB, and would still allow a high conversion rate. The noise in any stage of the pipelined ADC includes the accumulated thermal noise from all the previous stages [21, 22].

4.6.4 The effect of op-amp parameters in ADC

The op-amp is one of the most important blocks in the pipelined ADC. Its performance directly affects the SHA and MDAC transfer functions and their output settling voltages, which could lead to a high level of non-linearity errors in the pipelined ADC output. It is required to determine the maximum DNL error output from the pipelined ADC due to the op-amp finite open-loop gain and input parasitic capacitance. It is the source of maximum error.

4.6.5 Mismatch error of the capacitor

The linearity of a pipelined ADC is basically degraded by the linearity errors in its pipeline stages. Compared to gain errors, capacitor-mismatch errors have a significantly more degrading effect on the overall linearity of the pipeline stage [24].

The gain of a switched capacitor MDAC is determined by capacitor ratios. If the capacitors are not equal, then an error proportional to the mismatch is generated in the residue output. Thus, accurate capacitor matching is required to design a high resolution pipelined ADC. The capacitors' value is given by

$$C = A\frac{\varepsilon_{\text{ox}}}{t_{\text{ox}}} = AC_{\text{ox}}, \tag{4.18}$$

where A is the area of a capacitor, ε_{ox} is the dielectric constant of silicon dioxide, t_{ox} is the thickness of the oxide and C_{ox} is capacitance per unit area.

The capacitance value depends on the area and oxide thickness of a capacitor. The main causes of capacitor mismatch are due to wrong-etching and the variation of oxide thickness. Since C_{ox} is fixed by a process technology, the accuracy of capacitance can be improved by simply increasing the area. However, in SC circuits the accuracy of a capacitor ratio is more important than the absolute of a capacitor.

The integrated circuit capacitor can be defined as

$$C' = C + \Delta C,$$

where ΔC is the mismatch error of capacitor C. If the two capacitors are C_1 and C_2, the ratio of the two capacitors with mismatch can be expressed as

$$\frac{C_1 + \Delta C_1}{C_2 + \Delta C_2}. \tag{4.19}$$

The accuracy of the capacitor ratio can be improved if the difference of the mismatch errors of both capacitors is small. A mismatch error in the accuracy of a capacitor ratio due to over-etching can be minimized by implementing capacitors with an array of small equal sized unit capacitors. A mismatch error of capacitor ratios due to the variation of oxide thickness can be minimized by laying out the capacitors very close to each other.

4.6.6 Offset error of the comparator

A comparator generates an output signal indicating whether or not an input signal is larger than a reference one. When the comparator finds the difference between two input signals, an internal offset voltage is added to this difference [20, 25]. Thus, when the two inputs are close to each other, the comparator may make a wrong decision. When the comparator makes an incorrect decision, the output code is wrong and the incorrect reference is subtracted from the input. The result is the residue that is out of range of the next stage of the pipeline after amplification. The offset error of the sub-ADC is also the common error, but its effect will be nullified while employing the digital correction technique. At best 25% of the full scale range offset error can be eliminated with digital correction. The offset error was introduced by putting a dc voltage source of 50 mV magnitude at the negative input terminal of the dynamic comparator.

4.6.7 Gain error and offset error of the op-amp

The op-amps are the most important building blocks in analog circuits and also discrete data domain circuits, such as switched capacitor circuits which are needed for building some blocks of pipelined ADCs. Therefore, it is necessary to study the impact of the non-idealities of op-amps on pipelined ADCs. The op-amp has input parasitic capacitance C_P. Let the open-loop dc gain of the op-amp be A_0. During the sampling phase, the sampling capacitor C_S and the feedback capacitor C_F are connected into an input, sampling an input signal on the capacitors. During the amplification, the feedback capacitor C_F is connected at the output of the op-amp and the sampling capacitor C_S is connected with $\pm V_{\text{ref}}$ or to ground depending on the output of a sub-ADC. Error due to finite gain of the op-amp should be smaller than 1/2 LSB of the remaining resolution. The gain can be found from

$$\frac{1}{A_0\beta} < \frac{1}{2}\text{LSB}, \tag{4.20}$$

where the feedback factor is given by

$$\beta = \left(\frac{C_S}{C_S + C_F + C_p}\right).$$

The gain of the OTA which is used in the ADC stages for multiplying and digital-to-analog conversion operation has been decreased to create the gain error, since with changes in supply voltage and temperatures the gain will worsen in submicron technologies, and the offset error for the OTA has been created again by a dc source of 50 mV magnitude. All of these four types of faults are created and simulated individually to know their individual behaviour in the system. It is found that the result is quite satisfactory.

4.7 Preparation of the layout and post-layout simulations

4.7.1 Preparation of the layout of the whole system

The layout for the entire BIST system has been prepared by utilizing four metal layers and a poly-silicon layer in the CMOS9T standard 180 nm technology process. While preparing the layout each transistor has been prepared from the fundamentals by making all individual layers instead of using templates of the transistors which are designed for particular technologies, such that the area occupied by the system is optimized and all the lengths, widths and spacings between similar layers are maintained as the minimum dimensions allowed by the technology. Maintaining the minimum dimensions, the parasitic resistances and capacitances will also be reduced and hence better performance will be assured.

The total chip including BIST and CUT has an area of 534 μm × 312 μm = 0.1666 mm^2, whereas the CUT, i.e. the pipelined ADC, occupies 454 μm × 280 μm = 0.1271 mm^2. The testing circuit and the CUT occupy 23.7% and 76.3% of the total area, respectively, as shown in figure 4.36. The testing circuit consists of the stimulus generator and analog multiplexers and requires clock generation to operate the multiplexers for switching between the actual running mode and testing mode. An output response analyser (ORA) has not been designed, and the spectre simulator is used as the ORA. After preparing the layout, post-layout simulations are also performed to verify its functionality once again. From the post-layout simulations it is observed that because of parasitic elements, there is a slight variation in the functionality of the system. The occupied area of the major blocks is shown in table 4.1, and also the post-layout simulations of the entire system and the OTA block which is used for the filter and VCO are shown in figures 4.37 and 4.38.

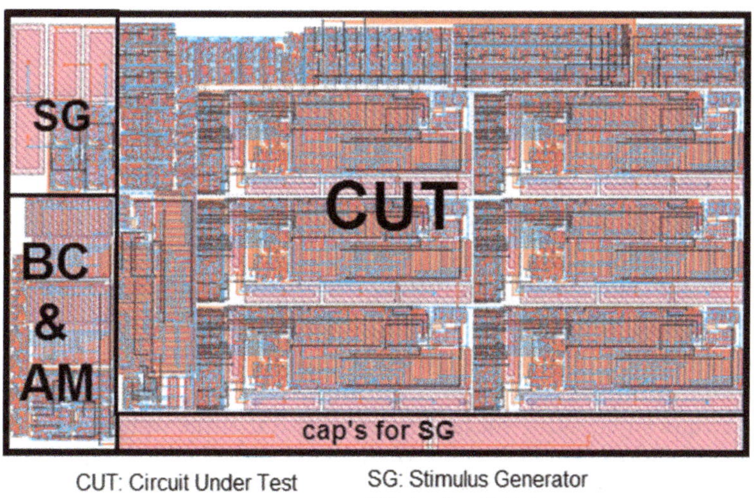

Figure 4.36. Layout for the total BIST system.

Table 4.1. Block-wise distribution of the layout area.

Major blocks	Area occupied (μm × μm)
Folded cascode OTA	59.8 × 131.4
S&H circuit	59.8 × 156.6
MDAC	74 × 176
Pipeline stage	74 × 197.5
Digital error correction	37.2 × 92.4
Time alignment	45.4 × 213.1
Thermal to binary converter	14.9 × 126.6
OTA-C filter	78.4 × 119
Total tuning system	183 × 129.8
PLL	51.7 × 63.7

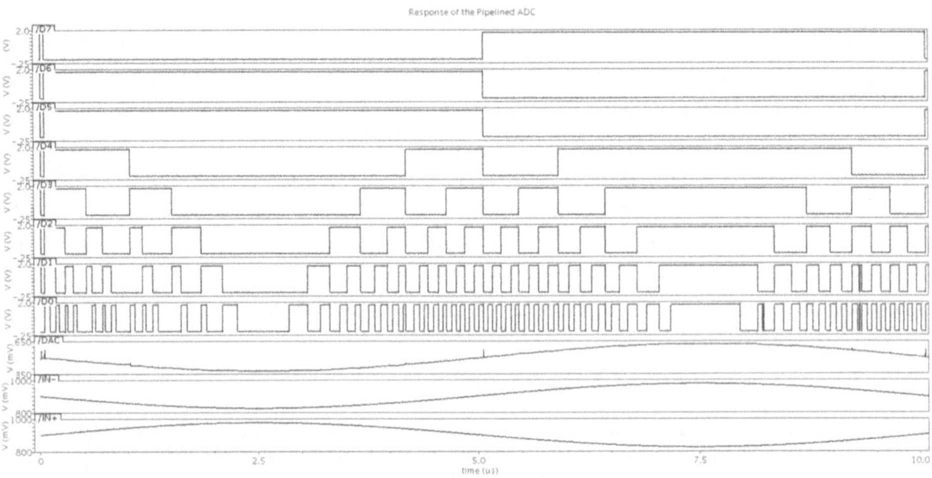

Figure 4.37. Post-layout simulation result of the pipelined ADC with a 100 KHz frequency input signal.

4.8 Conclusion

The work presented in this chapter was performed in the Cadence environment with the CMOS9T process, 0.18 μm technology. To perform dynamic performance testing through the oscillation-based BIST, for the 1.8 V, 100 MSPS pipelined ADC, a 1 V_{p-p} signal with a frequency of 1 KHz to 1 MHz is required as a test stimulus. To generate the on-chip test stimulus a new approach has been used to generate the sine wave from the square wave by using a special non-linear wave shaping circuit, which removes the unwanted harmonics from the spectrum of the square wave, leaving few higher-order harmonics, with a much smaller amount of silicon overhead on the test system and the main circuit which is to be tested. These remaining harmonics are further removed from the square wave spectrum by the analog filter which has been designed. Various types of filters are tested to work with the wave shaping circuit.

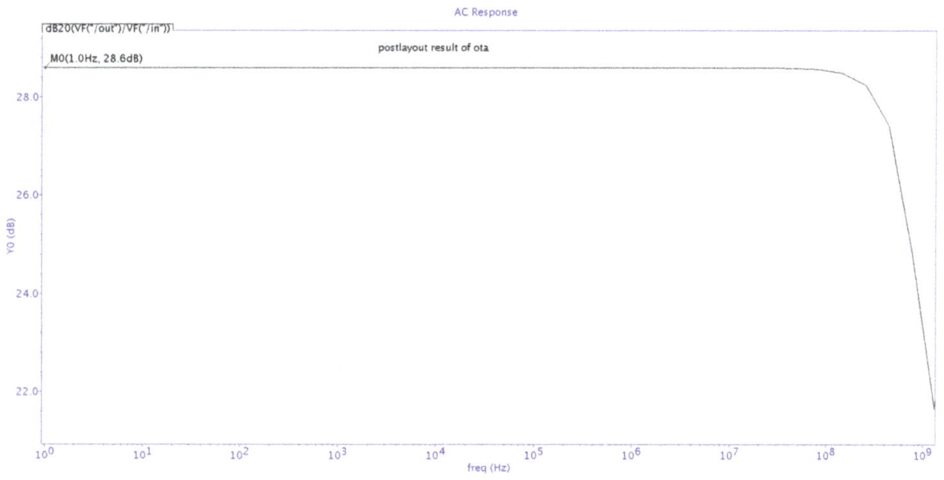

Figure 4.38. Post-layout plot showing the gain of the OTA which is used in the OTA-C filter.

Only the operational transconductance–capacitor filter gives the best results as a filter. The OTA-C filters are small in size and hence they are much prone to parasitic effects. In order to reduce the parasitic effects, an on-chip automatic tuning system is used to vary the transconductance values of the filter. The tuning circuit is designed from the basic techniques with a new approach. The test stimulus is generated with the help of this wave shaping circuit and OTA-C filter along with the automatic tuning system.

The generated test stimulus has been applied to the circuit under test which is a pipelined ADC. The output response of the ADC for the test stimulus is processed through a DAC to reconstruct the digital code into analog signal with a 2 bit greater resolution. Then a 1024 point fast Fourier transform has been applied on the reconstructed signal, from which the dynamic performance parameters of the pipelined ADC, such as the SNR, signal-to-noise plus distortion ratio, total harmonic distortion spurious free dynamic range and effective number of bits, have been evaluated and these values are stored as a good system response to differentiate between a healthy system and a faulty system.

Some of the most commonly occurring faults in the pipelined ADC were simulated manually by creating the fault environment in the ADC. Once again the dynamic performance parameters were evaluated for the faulty system to check the fault coverage of the system as well as to determine the effect of those faults on the dynamic performance parameters. These values were compared with the values of the dynamic parameters of the healthy system to determine whether the system is faulty or not.

The layout for the total system including the testing circuit and the circuit under test, i.e. the pipelined ADC, has been prepared and post-layout simulations were also carried on the layouts of all the circuits of the BIST system in the CMOS9T 0.18 μm 1P4M CMOS technology. The post-layout simulation shows that there is an effect of parasitic elements on the system, which is responsible for malfunction of the system. The total chip, including the circuit under test and the testing system, occupies an area of 0.1667 mm^2 on the wafer, whereas the CUT has 76.3% of the total chip area, i.e. 0.1271 mm^2, and its associated testing system has a 26.7% area. This is achieved with low power consumption. The pipelined ADC consumes 16.38 mW power and the testing circuit consumes only 1.076 mW.

Thus an oscillation-based BIST system for the 8 bit pipelined ADC with a sampling rate of 100 mega samples per second is designed by satisfying all the requirements for the self-testing system.

4.8.1 Future scope

The static performance parameters such as INL and DNL are to be found from the frequency response, and the effect of static performance parameters on the dynamic performance parameters needs to be understood, i.e. how the static performance parameters can affect the dynamic performance parameters of the pipelined ADC needs to be studied.

The time domain dynamic performance parameters, such as aperture delay, aperture jitter and over-voltage recovery, etc, have to be calculated from the BIST system.

References

[1] Lee D, Yoo K, Kim K, Han G and Kang S 2004 Code-width testing-based compact ADC BIST circuit *IEEE Trans. Circuits Syst.-II* **51** 603–6

[2] Barragán M J, Vázquez D and Rueda A 2007 Practical implementation of sine wave generators for mixed-signal BIST *Proc. of the 8th Latin American Test Workshop (LATW'07) (Cúzco, Peru)*

[3] Mrak P, Biasizzo A and Novac F 2010 Measurement accuracy of oscillation based test of analog to digital converters *ETRI J.* **32** 154–6

[4] Adamo F, Attivissimo F, Giaquinto N and Savino M 2002 FFT test of A/D converters to determine the integral non-linearity *IEEE Trans. Instrum. Meas.* **51** 1050–4

[5] Razavi B 2002 *Design of Analog CMOS Integrated Circuits* (New York: McGraw-Hill)

[6] Allen P E and Holberg D R 2002 *CMOS Analog Circuit Design* (Oxford: Oxford University Press)

[7] Papež V and Papežová S 2010 Sine-wave signal sources for dynamic ADC testing *Proc. 27th Int. Conf. on Microelectronics (Niš, Serbia)* pp 16–9

[8] Kabisathpathy P, Barua A and Sinha S 2005 *Fault Diagnosis of Analog Integrated Circuits* (Dordrecht: Springer)

[9] Schaumann R and Valkenburg M E V 2001 *Design of Analog Filters* (New York: Oxford University Press)

[10] Schaumann R, Ghausi M S and Laker K R 1981 *Design of Analog Filters—Passive, Active RC and Switched Capacitor* (Englewood Cliffs, NJ: Prentice Hall)

[11] Lo T-Y and Hung C-C 2007 A wide tuning range G_m–C continuous-time analog filter *IEEE Trans. Circuits Syst.—I* **54** 713–22

[12] Veeravalli A, Sánchez-Sinencio E and Silva-Martínez J 2002 Transconductance amplifier structures with very small transconductances: a comparative design approach *IEEE J. Solid-State Circuits* **37** 770–5

[13] Koziel S and Szczepanski S 2002 Design of highly linear tunable CMOS OTA for continuous-time filters *IEEE Trans. Circuits Syst.—II* **49** 110–22
[14] Wu Z, Rui F, Zhi-Yong Z and Wei-Dong C 2008 Design of a rail-to-rail constant-G_m CMOS operational amplifier *2009 World Congress on Computer Science and Information Engineering* (Los Alamitos, CA: IEEE Computer Society), pp 198–201
[15] Sümesağlam T and Karşılayan A İ 2003 A digital approach for automatic tuning of continuous-time high-Q filters *IEEE Trans. Circuits Syst.: Analog Digit. Signal Process.* **50** 755–61
[16] Eid E-S and El-Dib H 2009 Design of an 8-bit pipelined ADC with lower than 0.5 LSB DNL and INL without calibration *4th Int. Design and Test Workshop (IDT 2009)*
[17] Yuan J, Fung S W, Chan K Y and Xu R 2012 A 12-bit 20 MS/s 56.3 mW pipelined ADC with interpolation-based nonlinear calibration *IEEE Trans. Circuits Syst.—I* **59** 555–65
[18] Centurelli F, Monsurrò P and Trifiletti A 2010 Behavioural modelling for calibration of pipeline analog-to-digital converters *IEEE Trans. Circuits Syst.—I* **57** 1255–64
[19] Benetazzo L, Narduzzi C, Offelli C and Petri D 1992 A/D converter performance analysis by a frequency-domain approach *IEEE Trans. Instrum. Meas.* **41** 834–9
[20] Huang J-L, Huang X-L and Kang P-Y 2009 A self-testing assisted pipelined-ADC calibration technique *ASICON '09. IEEE 8th Int. Conf. on ASIC* pp 565–8
[21] Kledrowetz V and Haze J 2010 Analysis of non-ideal effects of pipelined ADC by using MATLAB-Simulink *Proc. Advances in Sensors, Signals and Materials* pp 85–8
[22] Mancini E, Rapuano S and Dallet D 2008 *A Distributed Test System for Pipelined ADCs* (Amsterdam: Elsevier)
[23] Taherzadeh-Sani M and Hamoui A A 2006 Digital background calibration of capacitor-mismatch errors in pipelined ADCs *IEEE Trans. Circuits Syst.—II* **53** 966–70
[24] Ding L, Sin S-W, Seng-Pan U and Martins R P 2010 An efficient DAC and inter-stage gain error calibration technique for multi-bit pipelined ADCs *IEEE Asia Pacific Conf. on Circuits and Systems (APCCAS)* pp 208–11
[25] Galton I 2000 Digital cancellation of D/A converter noise in pipelined A/D converters *IEEE Trans. Circuits Syst.—II* **47** 185–96
[26] Tausiff D M D 2010 A built-in self test on 1.8 V 8-bit 125 MSPS pipeline ADC *MTech Dissertation* Department of Electrical Engineering, IIT Kharagpur
[27] Barua A and Tausiff M 2011 A code width built-in-self-test circuit for 8 bit pipelined ADC *Proc. of 21st Int. Conf. on System Engineering (Las Vegas, NV, 16–18 August 2011)* pp 287–91
[28] Biswas A 2011 Design of oscillation based built in self test (BIST) system for 1.8 V 8-bit 125-mega samples per second pipeline ADC *MTech Dissertation* Department of Electrical Engineering, IIT Kharagpur
[29] Dhanunjay N 2012 An oscillation based built in self test (BIST) system for dynamic performance parameter evaluation of a 8 bit 100 MSPS pipelined ADC *MTech Dissertation* Department of Electrical Engineering, IIT Kharagpur
[30] Barua A and Dhanunjay N 2013 A built in self test system for dynamic performance parameter evaluation of pipelined analog to digital converter *Proc. of Int. Conf. on Circuits and Systems for World Congress on Engineering and Computer Science (San Francisco, CA, 23–25 October 2013)* pp 671–6
[31] Barua A 2014 *Analog Signal Processing: Analysis and Synthesis* (New Delhi: Wiley)

IOP Publishing

Chapter 5

A reconfigurable built-in self-test architecture for a pipelined ADC

Raghavendra Singh Raghava and Alok Barua

This chapter describes a new built-in self-test (BIST) technique suitable for both functional and structural testing of analog and mixed-signal circuits based on the oscillation-test methodology. An analog-to-digital converter (ADC) is used as a test vehicle to demonstrate the capability of the proposed OBIST technique for both functional and structural testing. In the BIST structure, the input signal of the pipelined ADC is made to oscillate linearly between two transition voltages and subsequently around a particular output code of the ADC. By counting the number of clock cycles during the rise or fall time of the oscillation and comparing this with the ideal number of clock periods, the differential non-linearity (DNL) and integral non-linearity (INL) error can be obtained. Diagnosis of the faulty block in the ADC up to an accuracy of one block is also carried out through this technique. An on-chip reference code generator circuit is made which sequentially changes the reference code, for which the ADC nonlinearities are measured. Simulation results using an 8 bit pipelined ADC designed using CMOS 180 nm technology in the Cadence Virtuoso environment are presented.

5.1 Introduction

5.1.1 Introduction to the built-in self-test (BIST)

In very large-scale integration (VLSI), domain testing is a very important field. Whether it is carried out in the design phase of the integrated circuit (IC) chip lifecycle for detection and identification of design errors or in the manufacturing stage to find manufacturing defects, at both levels it plays a crucial role. Even after this its role does not end, as during system operation testing targets continues to

doi:10.1088/978-0-7503-1732-0ch5

detect any fault sustained during operation that will produce incorrect or unwanted outcomes. Therefore, to guarantee delivery of the best possible products to buyers, versatile testing methods are implemented during chip design. However, there are two constraints: the cost of the testing methodology and the time taken for testing restrict the productivity [1, 2].

As technology advances, more complex ICs are being produced and they require more advanced and evolved test equipment to meet the desired criteria. Also, automatic test equipment is becoming more expensive as it follows the level of performance. Even if we do not consider the cost, the time needed in the sequential testing of ICs limits the production. Thus there is a need for a better testing technique which can reduce the production costs and time for complex ICs.

In addition to all this in high complexity mixed-signal circuits, there are many limitations on accessibility to the different components of the circuit under test (CUT) and also there is a serious restriction on the number of test pins. Several design-for-testability (DFT) techniques have been developed to increase the observability, and to minimize the number of test pins. The best solution is to use the existing pins of the chip for testing without having any additional pins.

The alternative approach is the built-in self-test (BIST) in which the control, test stimulus generator and measurement circuitry are placed on the same chip and should be able to present a binary pass or fail result. A BIST solution, if proven, is the most promising test technique because it allows eliminating expensive mixed-signal testers and reducing the test time [3]. However, BIST should consume only a small percentage of overall chip area.

ADCs are a very important component of mixed-signal circuits. The design of high performance ADCs is today a very important research topic. Hence new test algorithms should also be developed to test high performance ADCs. Testing the ADCs has been identified as one of the major challenges for the future and BIST has been identified as one of the potential solutions to this testing problem.

Three different types of faults are considered and they are as follows:

- Static faults.
- Dynamic faults.
- Catastrophic faults.

In static fault coverage a low frequency signal is applied as the input to the ADC under test and the following parameters are measured from the digital output response.

5.1.1.1 Offset error

Offset error corresponds to the difference between the first practical transition voltage level and the first ideal transition voltage level normalized to the ideal least significant bit (LSB) size. It is shown in figure 5.1 with an offset error of + $1\frac{1}{2}$ LSB.

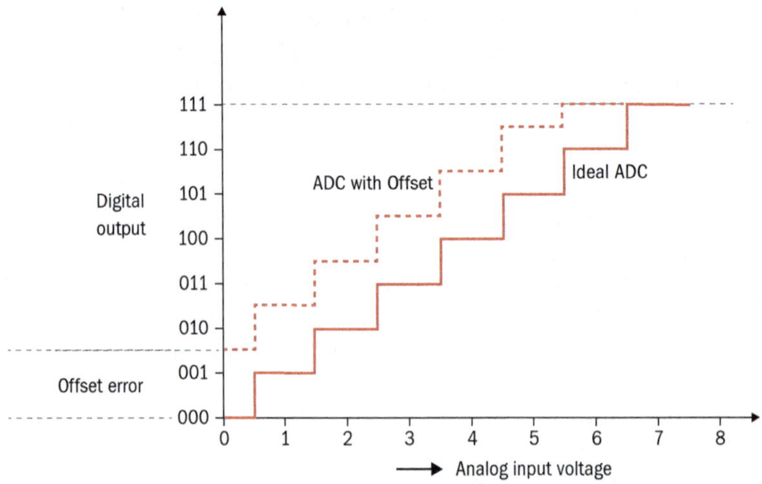

Figure 5.1. Offset error of an ADC.

5.1.1.2 Differential non-linearity (DNL)

DNL corresponds to the difference between the actual code width and ideal code width normalized in comparison to the ideal LSB code width (figure 5.2).

5.1.1.3 Integral non-linearity (INL)

INL corresponds to the maximum difference between the ideal and actual transition levels normalized with the ideal LSB size (figure 5.3).

5.1.1.4 Gain error

Gain error corresponds to the difference between the ideal slope between the zero and full scale code and the actual slope between the measured zero and full scale code.

5.1.1.5 Missing code error

Whenever the input to an ADC is increased by 1 LSB, if the difference between the two continuous output codes is greater than 1 LSB, then the ADC is said to have a missing code error.

5.1.1.6 Monotonicity error

When the input to an ADC is monotonically increased and if there is any decrease in the output code value, then the ADC is said to be faulty through monotonicity error. In dynamic fault testing of ADCs, a high frequency sinusoidal signal is generally applied as the input to the ADC and from the output response the following parameters are measured:

1. Signal-to-noise ratio (SNR).

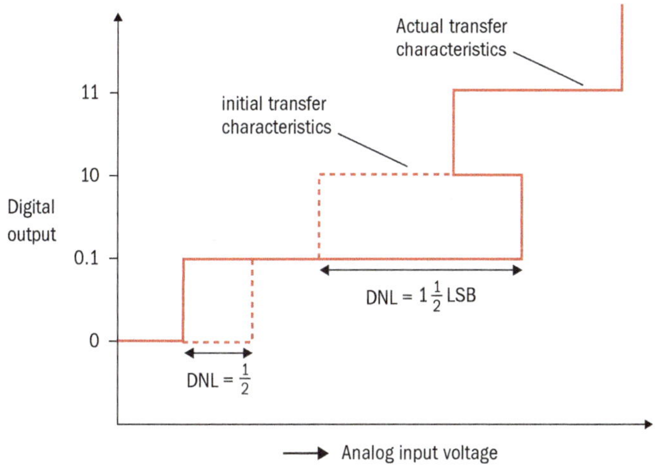

Figure 5.2. Output characteristic of an ADC showing DNL error.

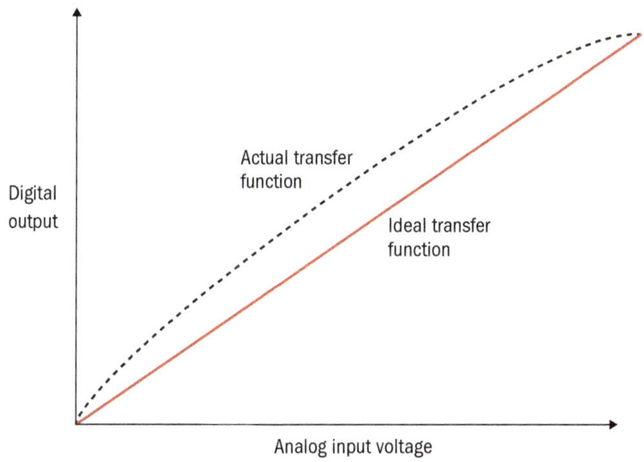

Figure 5.3. Output characteristic of an ADC showing INL error.

2. Total harmonic distortion (THD).
3. Spurious free dynamic range (SFDR).
4. Effective number of bits (ENOB).

In this work a reconfigurable oscillation-based BIST system has been designed for the static testing of a 1.8 V, 8 bit, 40 MSPS pipelined ADC. In this test scheme the ADC under test is made to oscillate around an output code and from the frequency of oscillation the static performance parameters are evaluated. Test stimulus generator and response analyzer circuits are designed. A 'control logic' block is made which controls the total testing process.

5.1.2 Literature review

A literature review of pipelined ADCs and BISTs is first carried out. For static testing of ADCs, the histogram test technique is a popular method [1–3]. It involves the application of an analog signal to the ADC input and recording the number of times each code appears at the ADC output. The analog signal can be any function whose amplitude distribution is known. ADC errors modify the output code count and hence the histogram shape. Hence measuring the output code count and comparing it to the ideal count and performing some complex calculations permits one to evaluate the offset error, gain error, DNL and INL. A triangular or sinusoidal signal is given as the input stimulus to the ADC. However, this BIST scheme requires a large amount of hardware resources.

Hardware minimization of histogram BIST for sinusoidal input signal has been carried out in [4, 5]. However, in the histogram BIST scheme the test time is a major issue for high resolution ADCs.

A BIST structure is proposed in [6] which permits one to evaluate the converter linearity only by using the LSB, the global functionality being tested with the comparison between the remaining bits and a counter clocked by the LSB. In a code-width-based static testing scheme a slope calibrated input signal is not required. The test circuit is a simple digital circuit with a gate count of only 550 and the simulated fault coverage is approximately 99% [7].

A fully digital test approach is proposed in [8]. Here the main advantage of the BIST circuit is that it can test the DNL and INL for all codes in the digital domain, which in turn eliminates the necessity for calibration. Again here some parts of the ADC under test are used in the BIST with minor modifications, which reduces the chip area and also the cost of the test. The proposed BIST structure presents a compromise between test accuracy, area overhead and test cost. In another BIST scheme, a delta-sigma-based linear ramp generation is proposed and applied to the ADC under test and the DNL and INL of the converter are measured [9]. Here 5% LSB test accuracy can be achieved in the presence of reasonable analog imperfection. The oscillation-based BIST scheme has been proposed by some researchers [9, 10]. The main advantage of this scheme is that it does not require an on-chip ADC or DAC, which makes it feasible for most mixed-signal ICs. Again, both the stimulus generation and measurement techniques are highly tolerant of analog variations. Based on sequential code analysis, another BIST scheme is proposed in [11]. In addition to the measurement of DNL and INL, non-monotonic behavior of an ADC can also be detected. A ramp generator is used to generate the test stimulus. Feedback configuration is used in the implementation of the ramp generator to increase linearity.

A low cost BIST solution for static and dynamic testing of an ADC is provided in [12]. Here the noise signal has been given as the test stimulus. The silicon area requirement and test time are also reduced in comparison to the histogram method. In a separate method, the dynamic performance of ADCs (THD and SFDR) is estimated from the INL data achieved from static characterization of an ADC, without requiring additional data acquisition or accurate sinusoidal sources [13, 14].

5.1.3 Motivation for this work

The oscillation-based BIST methodology is a general test scheme applied to functional and structural testing of mixed-signal circuits. The basic principle of this methodology is that the circuit under test is reconfigured to oscillate and important information regarding the functional and structural parameters of the circuit is extracted from the oscillation frequency. The deviation of the frequency of oscillation from its nominal value indicates a faulty circuit.

The oscillation-based test strategy has been successfully applied to a wide range of analog and mixed-signal circuits including analog filters and operational amplifiers. However, the application of this test strategy has never been used for functional testing of pipelined ADCs.

With the motivation above, the objective of this work is to apply the oscillation-based BIST methodology for functional testing of pipelined ADCs. A 1.8 V, 8 bit, 40 MSPS pipelined ADC is used as the circuit under test (CUT) to test the OBIST methodology and subsequently to evaluate the DNL and INL errors.

Regarding the reconfigurability, an 8×1 analog multiplexer is designed which takes the input after every individual block of ADC sequentially and after evaluation diagnoses exactly which block is faulty.

5.1.4 Objective of the work

- The primary objective of the work is to design a reconfigurable oscillation built-in self-test architecture for an 8 bit, 40 MSPS pipelined ADC in CMOS technology (180 nm) that will verify static as well as catastrophic faults.
- The accuracy of the design will be validated against the behavior observed against an ideal 8 bit ADC model designed in Verilog AMS in the Cadence Virtuoso environment.
- Finally the layout of the architecture with pipelined ADC will be developed and optimized so that it will cover a minimum area.

5.1.5 Contributions to the work

The project started with the design of pipelined ADC of moderate specifications which are given in the next section. An 8 bit parallel pipelined ADC is designed in the Cadence environment using 180 nm technology. In the course of time, the main blocks of a pipelined ADC are designed, which include a dynamic comparator and a folded cascode operational transconductance amplifier. Then sample-and-hold (S&H) circuit is designed followed by individual 1 bit stages of ADC. After that the digital part of the ADC is designed, which consists of a delay correction circuit, digital correction circuit, etc. Finally all the blocks are assembled and the simulation is performed.

After the completion of the pipelined ADC, we focus on oscillation-based BIST (OBIST) blocks. We started with the current reference generator which generates a triangular wave around the desired transition voltage level, which forces the ADC to oscillate between two codes corresponding to the transition voltage level. The main specification is to keep the slope not so high that the ADC skips the code and not so low that it will take a lot of time.

After the design of the current reference generator, a frequency-to-number converter is designed which changes the oscillation frequency to a specific number which is then compared to the value of an ideal ADC and static parameters are evaluated.

An 8 × 1 analog multiplexer is then designed and implemented which allows taking the inputs from intermediate blocks of the ADC. Then the signal is fed to the frequency-to-number converter and then it is subtracted from the ideal value in an 8 bit subtractor which is also designed and implemented on-chip. After that we multiply the residual value by 100 through a 7 × 7 bit multiplier so that we can divide it easily by the ideal ADC value in an 8 × 8 bit divider. Hence the value obtained will give the exact DNL value at that code value. Also, this DNL value is then fed to an adder circuit which adds all the values of DNL up to that code to give the value of the INL.

5.1.6 Organization of the chapter

This chapter is organized as follows. Section 5.1 contains an introduction to the work, the literature review, and the motivation and contribution to the work. Section 5.2 describes the working and architecture of the pipelined ADC. This section contains all the information related to the different blocks of the ADC. Section 5.3 covers the BIST and oscillation BIST methodology in a very detailed version. All aspects of the methodology are discussed and OBIST blocks are dealt with in this section. Section 5.4 explains the circuit implementation of all the major blocks of the ADC and OBIST along with the intricacies and contains the design constraints. The simulation results for each block simulation are also presented and discussed in this section. In section 5.5, the conclusions and a brief summary of the work are presented. Finally, the future scope of the work is discussed.

5.2 The pipelined ADC

5.2.1 Background

ADCs provide a link between the analog and digital signals. Figure 5.4 shows a basic diagram of an ADC. A continuous time analog signal input is converted to a discrete digital signal at the output. This digital output can then be processed by a digital system such as a processor or an FPGA.

An ADC samples an analog waveform at uniform time intervals and assigns a digital value to each sample. This digital representation, given by the ratio of the

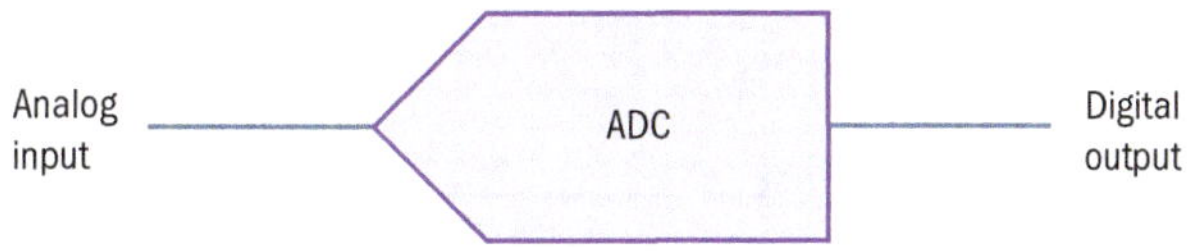

Figure 5.4. ADC principle.

actual voltage signal to a reference voltage, is an approximation of the original analog signal:

$$\text{DIGITAL CODE} = \frac{V_A}{V_{\text{ref}}} * (2^N - 1) = b_1 2^{-1} + b_2 2^{-2} + \cdots b_N 2^{-N} \ldots, \tag{5.1}$$

where V_A is the analog input voltage signal, V_{ref} is a reference voltage, N is the ADC resolution and b is the binary coefficient that has a value 0 or 1. Clearly the accuracy of an analog signal's digital representation is improved by using more bits at the ADC output.

Choosing a sampling frequency to ensure that the sampled signal contains sufficient information about the original signal and prevents aliasing, can be done based on the Nyquist–Shannon sampling theorem. That is, if the sampling frequency is more than double the signal frequency then the input signal can be fully recovered. This holds as long as the samples are not restricted to discrete values, as they are in a digital signal. The discrete nature of the values introduces errors due to quantization [12]. ADCs can be divided into two major categories based on their sampling frequencies: oversampling and Nyquist converters.

5.2.1.1 Oversampling converter

Oversampling converters are characterized by a sampling frequency much higher than the Nyquist rate. This high sampling rate causes larger spacing in the signal spectrum, ideally preventing the overlap of samples in the spectrum that leads to aliasing effects. These converters are typically used when high accuracy is required and a reduction in the effects of aliasing is desired, such as in band limited signals such as music. The design trade-off for the accuracy is a lower throughput. These converters also require a large number of samples to perform a single conversion.

5.2.1.2 Nyquist converters

Nyquist converters can process signals up to one half of the sampling frequency. These converters have higher throughput than oversampling converters. The trade-off made for this speed is reduced accuracy. Some Nyquist converters are high speed, with what is considered to be low to medium accuracy, such as flash or pipelined ADCs. Other Nyquist converters fall into the middle range for both speed and accuracy, such as successive approximation converters (SARs) and cyclic converters.

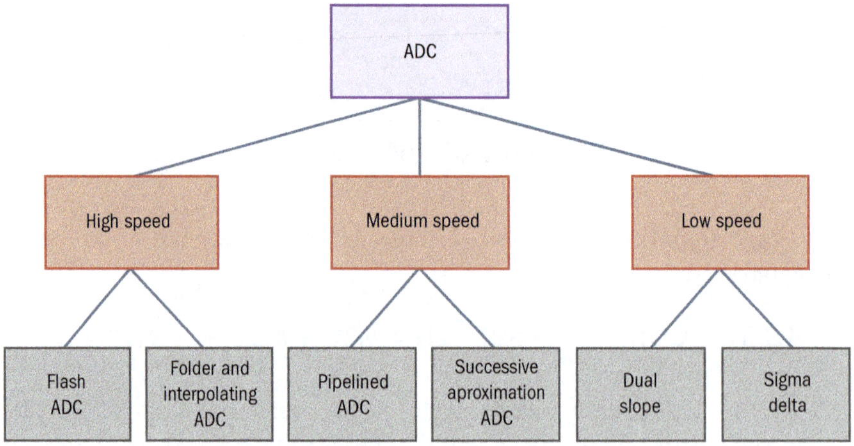

Figure 5.5. Classification of ADCs.

5.2.1.3 Classification of ADCs

ADCs are often divided into three major categories based on the converter speed and accuracy. This is shown in figure 5.5.

Flash ADC has low resolution, whereas folding and interpolating, pipelined, successive approximation and dual slope ADCs have moderate resolution. However, sigma-delta ADC has high resolution.

5.2.2 Introduction to pipelined ADCs

With the evolution of portable electronic devices, namely cell phones, tablets and laptops, low power, high resolution and high speed ADCs are becoming more and more necessary [11]. The categories of ADCs are numerous, each specialized for specific performance parameters. Modern converters have implemented, in parallel, processing of the analog signal to improve the bit resolution while still working at the same speed. Converters under this domain are the folding, multi-step and pipelined ADCs [12].

Generally ADCs are not strictly limited to pure flash, folding or pipelined. Hybrid structures with pipelined folding stages or smaller pipelined stages with a large flash section at the end are also options [12]. With a movement toward system-on-a-chip (SOC) high performance converters are frequently implemented on the same chip with microcontrollers and other digital systems. This introduces new noise and fabrication challenges which are not so dominant in discrete time domain converter implementations. Additionally, processes tailored for digital logic are not the best processes to make the analog circuits required for analog-to-digital conversion, but are becoming more frequently the place where converters physically take shape. With this movement toward chip level integration it is desirable to have a converter architecture that is tolerant of matching and process errors as well as noise introduced by adjacent devices [12]. With this in mind, this work focuses on the pipelined architecture because of its high tolerance of process variations, low power consumption and small area, making it an ideal candidate for system level integration in a mixed-signal circuit.

Most of pipelined ADCs are implemented in switched-capacitor (SC) circuits [1]. The performance of an SC implemented pipelined ADC is determined by the operational amplifier and the capacitor ratios. The op-amp must have a high dc gain, high slew rate and wide bandwidth to meet the accuracy and speed requirements. The capacitor ratio is another factor that limits the pipelined ADC performance. Therefore, care should be taken to achieve precise values of capacitor ratios during fabrication. The individual values of capacitors are not very important in SC circuits [15].

5.2.2.1 Pipelined ADC principles and architecture

Pipelined ADCs operate on a principle similar to that of the long division method. The quotient maps to the digital output bits and the remainder to the residue. An analogy to the above can be provided step by step. First, the dividend and divisor are taken. For example to divide 59 by 8:

```
7
-----------
8 | 59
56
---------
3
```

This is in the decimal system (base 10). Here 7 corresponds to the first bit and 3 corresponds to the residue. Multiply 3 by the base (10):

```
7.8
---------
8|59
56
--------
30
28
-----
2
```

The above procedure is carried out until the desired resolution is achieved. The number of times the above division is carried out corresponds to the number of stages in the pipelined ADC. One can see that a small error in the subtraction or multiplication could lead to gross errors at the output (in particular for high resolutions). Generally, $1\frac{1}{2}$ bits per stage pipelined architecture is employed for achieving better resolution and meets the design specification with a relatively easier design complexity. Here, 1.5 bits refers to the overlapping of one bit between adjacent stages (each of which would produce a 2 bit output). This added redundancy is popularly called digital correction [16, 17].

A pipelined ADC is inherently a multi-step amplitude quantizer in which the digitization is performed by a cascade of many topologically similar or identical stages of low resolution analog-to-digital encoders. Pipelining enables high conversion throughput by sample-and-hold amplifiers (SHAs), between stages that allow a concurrent operation of all stages. This is achieved at the cost of an increased latency. The block diagram of a pipelined ADC is shown in figure 5.6.

A pipelined stage takes two actions when an input signal arrives (signaled by a master clock) at the output of the SHA, a coarse quantization is performed by the sub-ADC. These two operations are often performed simultaneously. The resolution of the conversion is improved by passing a residue signal (the unconverted part of the input signal) to the later stages where it is further quantized. The conversion residue is created by a digital-to-analog converter (DAC) and a subtraction circuit. A typical pipelined ADC stage usually consists of a coarse comparator and a switched-capacitor circuit termed the multiplier DAC (MDAC), which integrates the S&H, the DAC, the subtraction and the residue-gain functions. This keeps the signal level constant and allows the sharing of an identical reference throughout the pipelined stages [13].

The number of bits/stage has a large impact on the speed, power and accuracy requirements of each stage, where high speed and low resolution correspond to having a low number of bits per stage and vice versa [2]. Therefore, in our design, in order to meet the specifications, $1\frac{1}{2}$ bits/stage were chosen, as shown in figure 5.7.

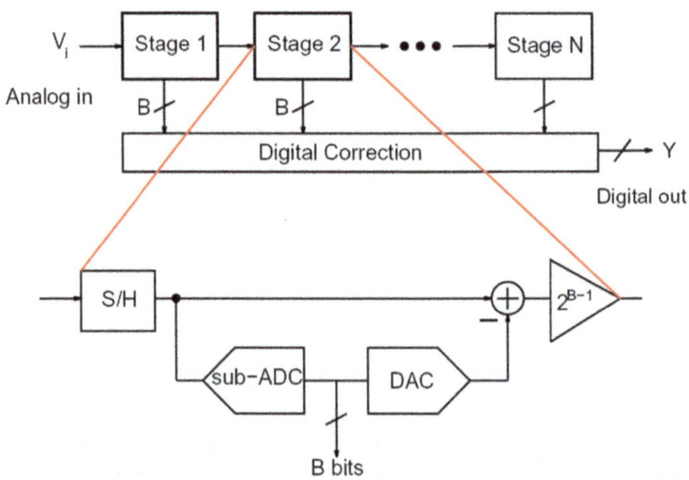

Figure 5.6. Block diagram of a pipelined ADC.

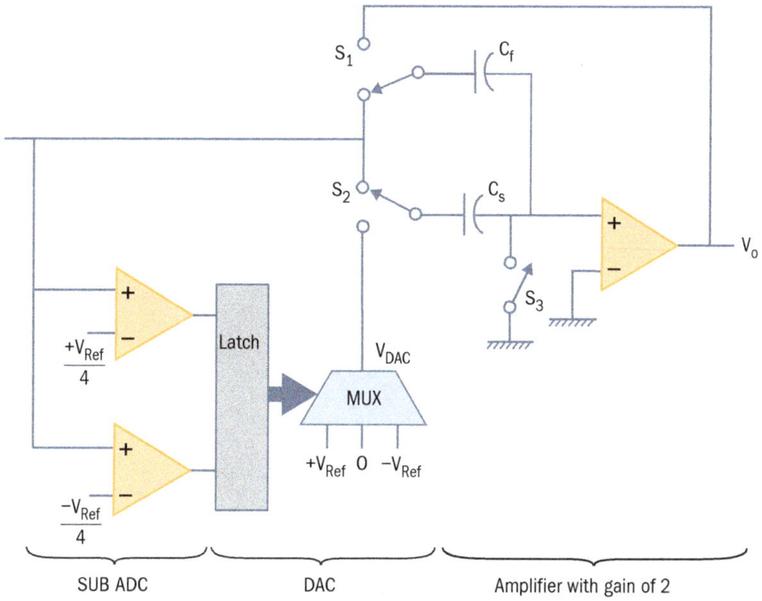

Figure 5.7. Architecture of single stage of $1\frac{1}{2}$ bits/stage ADC.

The basic idea was to have less constraint on the comparator and to match the bandwidth requirements.

The complete design consists of seven pipelined stages and one unity gain S&H circuit. The stages from 1 to 6 are identical and contain an ADC, a DAC, a subtractor and an amplifier configured for a gain of 2. Stage 7 is a specific stage containing no amplifier but only three comparators to save space and power consumption. The analog input signal is processed by an S&H block. In each successive stage, the input signal is then converted to digital format by an ADC section and the resulting 2 bits from the conversion are then applied to the DAC. Finally, the DAC output which can be either $V_{\text{ref}+}$, V_{CM} or $V_{\text{ref}-}$, is subtracted from the original input and the remainder is then amplified by the factor of two. The digital section of the ADC receives the outputs from the stages and aligns them with an appropriate delay. Then by simple addition the decisions are summed together in a $1\frac{1}{2}$ bits/stage method to give an 8 bit output word. All analog blocks are implemented in fully differential mode to have good power supply rejection ratio and also to decrease the influence of overall noise and even order harmonics [14].

5.2.2.2 S&H block

The S&H block is an integral part of each stage. This circuit has the input analog signal and at determined time intervals holds the voltage across a capacitor. The S&H circuit captures a sampled value of the continuous time input signals and the capacitor stores the sampled value which is a time invariant signal within that aperture. This stored signal is used by the other part of the circuit such as an ADC.

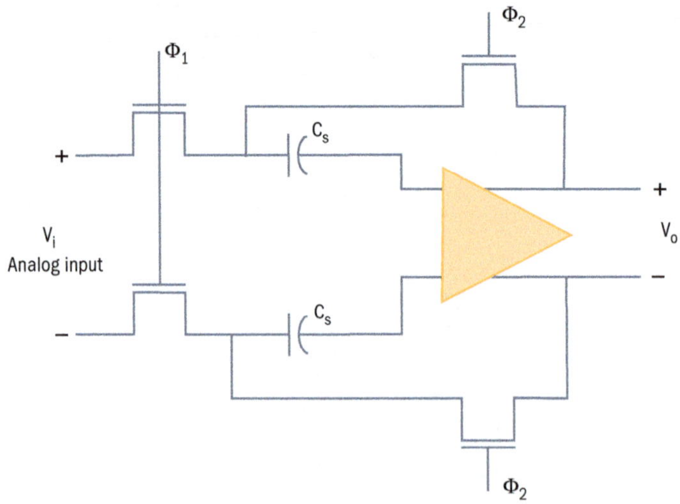

Figure 5.8. Block diagram of an S&H circuit.

It helps to smooth the fast changes of input signal and then relax the bandwidth requirements of the next parts. Figure 5.8 shows the schematic of the unity gain S&H that is used as the first stage in our design. It is a simplified model and in fully differential mode. This circuit uses a two-phase, non-overlapping clock. During phase Φ_1 the input V_i is sampled on the input capacitors (C_s), during phase Φ_2 the op-amp is put into a unity gain buffer configuration and the output is set to the sampled input value of $V_o = V_i$.

One limitation common to all data converters is the kT/C noise. It appears in S&H systems due to the thermal noise associated with the sampling switches. Obviously, the kT/C thermal noise only goes to zero for infinite sampling capacitance or absolute zero temperature. Thus kT/C thermal noise is a fundamental limit of any S&H circuit. The noise power in the base-band is given by

$$P_{n,}C_s = kT/C_s, \tag{5.2}$$

where C_s is the sampling capacitor. If the sampling capacitance increases, the noise voltage diminishes. There is a trade-off here for selecting the capacitor. A large capacitor improves the noise but requires more area and also greater power consumption.

5.2.2.3 Sub-ADCs and DACs

The quantization is performed by the sub-ADC block which resolves $1\frac{1}{2}$ bits of resolution and provides the necessary values for subtraction from the input value which, after being amplified by 2, results in the residue signal for the next stage. Each sub-ADC contains two comparators and one DAC.

The comparators decide on the value of the input voltage of each stage by comparing against two reference voltages $+V_{\text{ref}}$ and $-V_{\text{ref}}$ and then generate two outputs. Since the operation of the comparators is differential, the references can simply be inverted for the second comparator. Logic in the DAC section receives the output from the comparators and generates the control signal for the switches in the DAC block and the digital outputs to the delay nets [17].

The sub-ADC block determines the digital output of each stage and generates the decision signals that feed into the switched-capacitor network decision switches. Two comparators and digital logic are necessary to make a sub-ADC.

5.2.2.4 Residue amplifier

The purpose of the residue amplifier is to amplify the residue at each stage in order to scale it back to the range of the sub-ADC block. With the amplified residue value, the comparator in the sub-ADC of the next stage can make the right decision and acquire the correct digital value at the output of the sub-ADC. The residue amplifier is highlighted in figure 5.9. The input of the residue amplifier is the residue produced by the switched capacitor based on the comparator decision. This residue amplifier is fully differential with both differential inputs and differential outputs. Generally an operational transconductance amplifier (OTA) is used as a residual amplifier [18].

5.2.2.5 OTA architecture: a folded cascode OTA

In order to achieve high dc gain of an OTA as well as a high gain bandwidth product, the selection of the architecture of the OTA is of prime concern. The dc gain A_{dc} can be increased if a two-stage amplifier is considered. However, in such circuits, to achieve stability, compensation capacitors are required. These capacitors help form Miller effect capacitances which in turn increase the effective time constant of the amplifier. Thus the operational speed of the amplifier reduces. Cascoding a single-stage amplifier increases its output impedance and the dc gain A_{dc} according to the equation $A_{dc} = G_m \cdot R_{out}$. Also, a single-stage amplifier works much faster than its two-stage amplifier. However, cascode OTAs have limited output swing (OS) and input common mode range (ICMR). A folded cascode OTA architecture removes these limitations to some extent. In figure 5.10 a cascode OTA is compared to a folded cascode OTA. Comparing the ICMRs, it is found that the folded cascode OTA shows great improvement over the telescopic OTA. As far as dc gain is concerned, a folded cascode OTA has a comparable dc gain to that of a

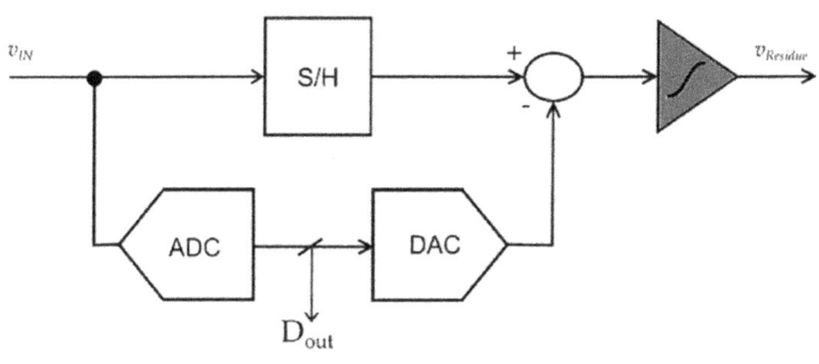

Figure 5.9. A single stage of an ADC showing the residual amplifier.

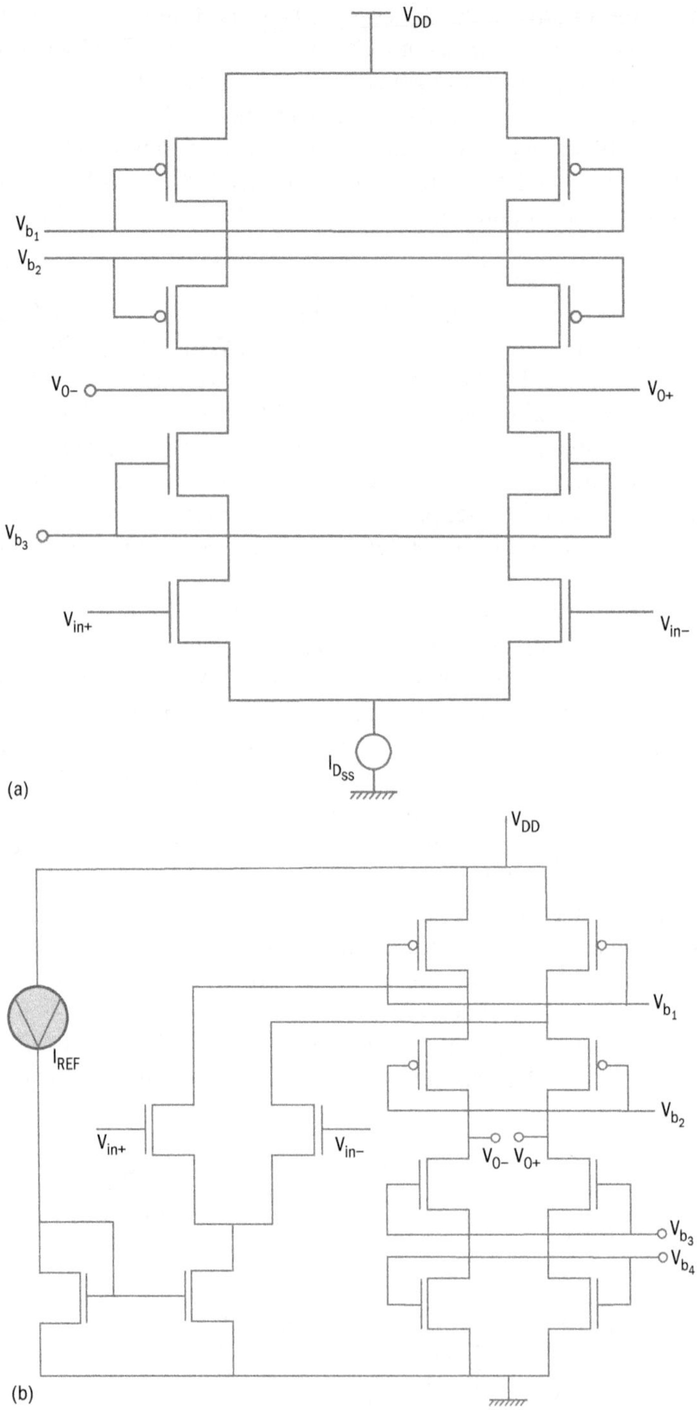

Figure 5.10. (a) Cascode amplifier. (b) Folded cascode amplifier.

cascode OTA. From figure 5.10 it is found that both OTAs have a transconductance of $g_{m_1m_2}$. Both OTAs also have an output impedance of the order of $g_m r_{ds}^2$ which will obviously provide a high dc gain [19].

5.2.2.6 The digital part

Errors caused by offsets in the comparator can be removed by using a digital correction technique. To what extent comparator errors can be canceled depends on how many bits each stage is designed to incorporate. In the case of a $1\frac{1}{2}$ bits per stage architecture comparator, offsets of up to 1/4 voltage full scale can be effectively handled.

An error occurs when the next stage cannot properly convert an out-of-range signal [5]. Without using digital correction the output bits are not useful. Digital correction contains two parts: the delay circuit and error correction. Because of the delays inherent in a pipelined architecture, most significant bits (MSBs) are provided a few clock cycles before the LSB, so as to align the bits in time a delay circuit is required. The delay network can be made of clocked D-latches since every successive stage is inversely clocked and there is no need to use flip-flops [20]. The digital delay correction circuit is shown in figure 5.11.

5.2.2.7 Clock circuit

The parallel pipelined ADC discussed in this work uses a switched-capacitor circuit. In such circuits, switches are closed or opened depending on the clock signal controlling their gate terminals. Thus, it is obvious that no two switches, which are supposed to work at alternate clock cycles, must be close together. To fulfill this requirement, the clock signals generated must be non-overlapping in nature [15]. A non-overlapping clock is shown in figure 5.12 with the top one the input clock signal in the non-overlapping clock generation system. The circuit schematic in Cadence is shown in figure 5.13.

5.3 Oscillation-based built-in self-test system

5.3.1 General BIST principles

The basic idea of BIST is to design a circuit so that it can test itself and determine whether it is faulty or fault-free. This typically requires that additional circuitry and functionality be incorporated into the design of the circuit to add self-testing features. This additional functionality must be capable of generating test patterns as well as providing a mechanism to determine if the output response of the circuit under test (CUT) to the test patterns corresponds to that of a fault-free circuit. Thus BIST is basically a method of testing an integrated circuit (IC) that uses special circuits built into the IC. This circuitry performs test functions on the IC and signals whether the different blocks of the IC covered by the BIST circuit are working properly [11].

The basic test flow of the BIST is shown in figure 5.14. A determined set of input stimuli created by a test pattern generator is applied to the circuit under test (CUT) and the output response of the circuit corresponding to the test stimuli is compared to a known healthy response or expected response to determine if the circuit is good

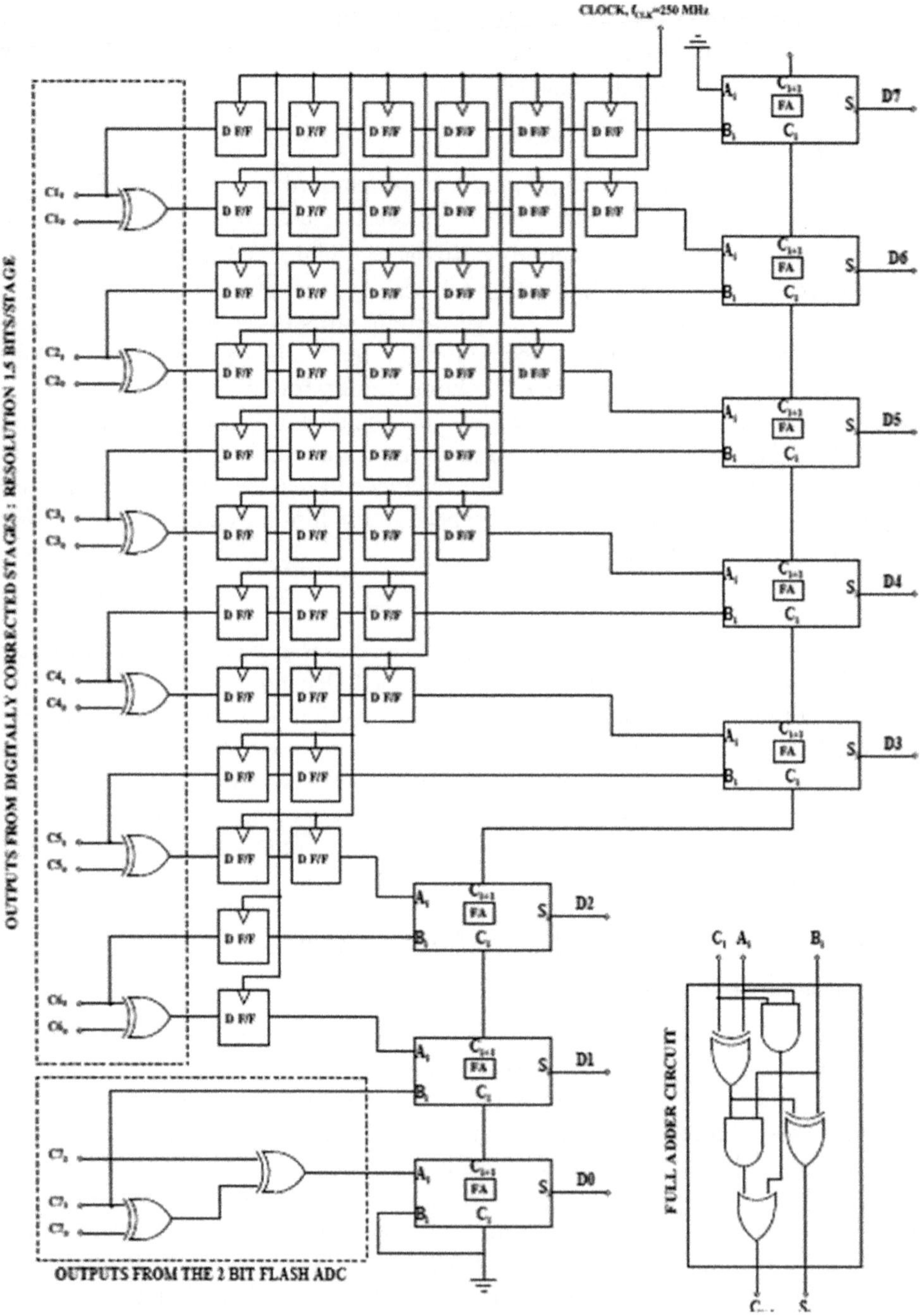

Figure 5.11. Digital and delay correction circuit.

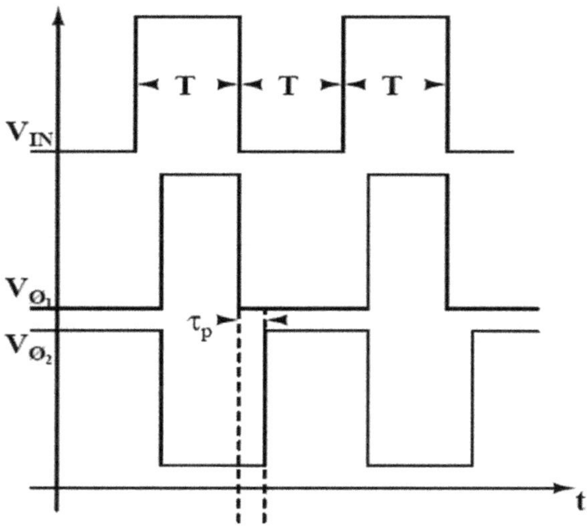

Figure 5.12. Output characteristic of the clock circuit.

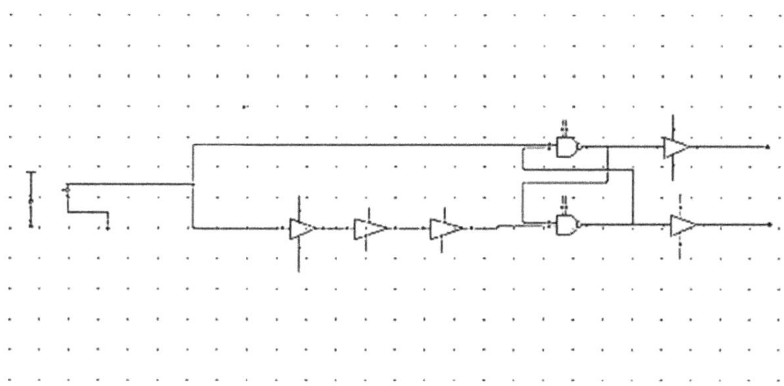

Figure 5.13. Schematic of the clock circuit in Cadence.

or faulty. There may be feature extractor circuits which extract important information from the output response and then compare them to the expected test parameters [21].

5.3.2 Oscillation-based built-in self-test (OBIST) system

The oscillation-based test method is basically a procedure of converting the CUT to a circuit which oscillates. The oscillation frequency, which is related to functional and/or structural parameters of the CUT, is evaluated. The deviation of the oscillation frequency from its nominal value indicates a faulty circuit. This test method does not require test vector generation or any input stimulus for testing. A generic block diagram of the proposed OBIST technique is illustrated in figure 5.15. The OBIST circuitry is composed of an analog multiplexer (AMUX), which selects

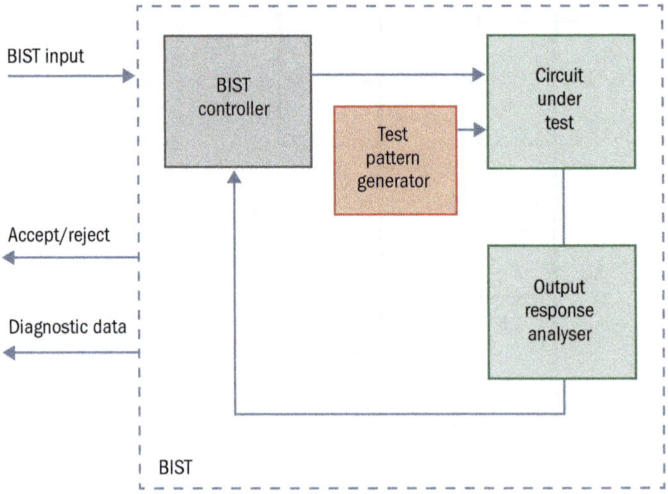

Figure 5.14. Architecture of a general BIST scheme.

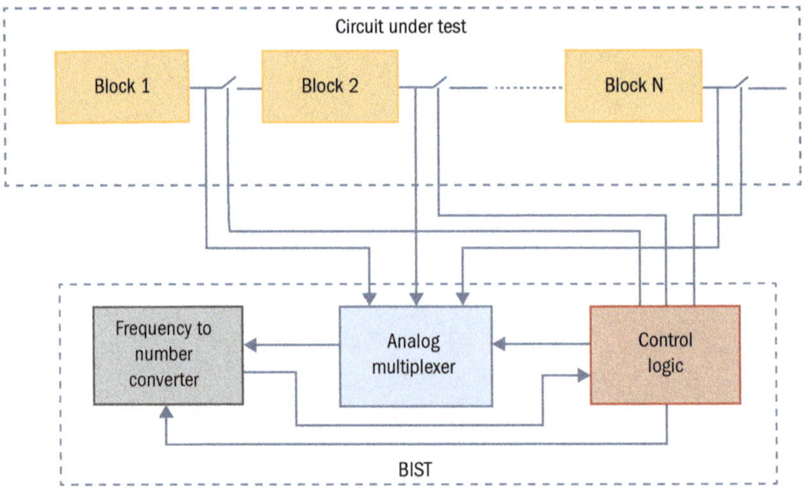

Figure 5.15. Architecture of the OBIST scheme.

the block of the CUT to be evaluated, a frequency-to-number converter (FNC), which converts the oscillation frequency selected by AMUX, and control logic, which directs all operations and produces a flag describing a healthy or defective block of the CUT [22].

In the test mode, the CUT is partitioned into appropriate blocks which are rearranged as oscillators one by one. An analog multiplexer (AMUX) selects the test point extracted from the converted building block. The oscillation frequency of the selected block is then converted to a number using the FNC, and is evaluated by the control logic.

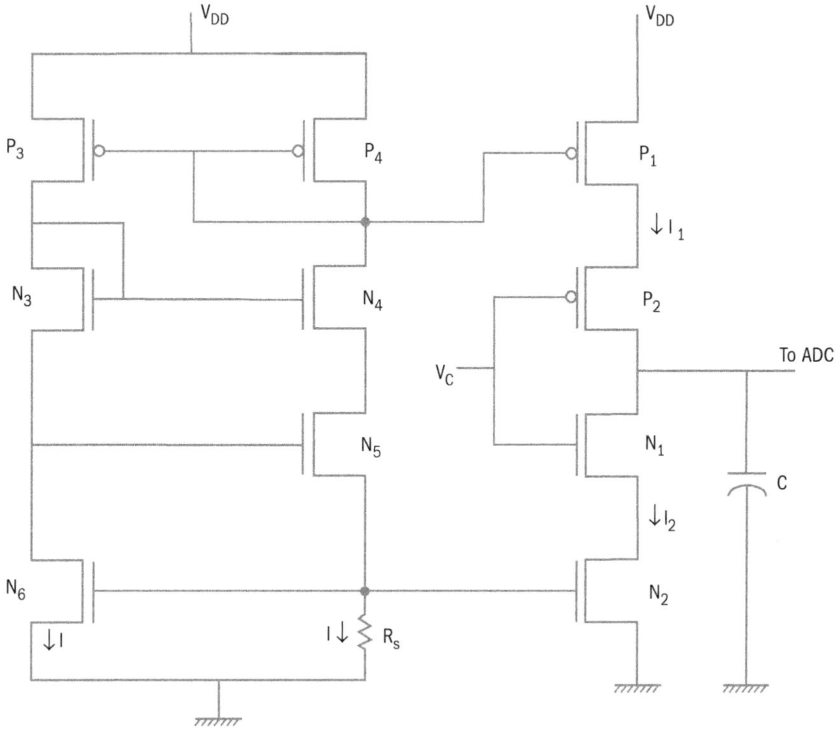

Figure 5.16. The current reference circuit.

5.3.3 Design of the current reference

This circuit produces a triangular wave around the threshold transition voltage of the ADC for a specific code when a switching frequency of value f_{osc} is given to the circuit [23, 24]. The current reference circuit is shown in figure 5.16.

The noise generated by the current source is very small in comparison to that of the ADC under test and therefore does not affect the measurements. The main error source due to hardware limitations is introduced by the mismatch error between the I_1 and I_2 currents, which causes a difference in the rise and fall of the input voltage of the ADC.

The voltage across a capacitor increases or decreases linearly when a capacitor is charged or discharged through a constant current source. The capacitor will be charged or discharged depending on the voltage V_c which is controlled by a control signal from the control logic block. If V_c is low (positive logic) the capacitor C will be charged via P_1 and P_2 and if it is high the capacitor will be discharged through N_1 and N_2. The transistors (P_1, P_2 and N_1, N_2) will be chosen properly to make currents I_1 and I_2 close to each other.

As shown in figure 5.16, where a signal generator is designed using CMOS 180 nm technology, a supply-voltage independent reference current of I_{ref} is set up using N_3–N_6, P_3, P_4 and resistor R_S. This current is mirrored in transistors N_2 and P_1 to produce I_1 (sink) and I_2 (source), respectively. In either case, the current mirrors copy the same reference current, thus ensuring good source/sink current matching. The absolute value of the reference current I_{ref} can be changed to control the currents I_1 and I_2. The main problem of this and any other implementation is the variation of I and C values due to process variation and chip temperature changes. The tolerances of I and C affect the oscillation frequency and introduce measurement error. This error can be tolerated in DNL testing but will affect the accuracy of the INL error test [9].

The design of the signal generator mainly involves the selection of the constant current source value and the capacitor value to control the slope of the ramp. The selection of the slope of the ramp is dependent on several factors which are as follows:

- The slope should be slow enough such that several samples are taken for a change of 1 LSB at the output. If the slope is made higher, then there is the chance that the output of the ADC misses some output code and delivers some output code which is greater than 1 LSB from the previous code. Hence a false missing code error will be shown. Further, for good resolution of the calculation in the error measurement blocks (offset error, DNL, INL and gain error), the slope of the ramp should be low.

  However, as we decrease the slope of the ramp, the test time increases.
- For a reference code A_j, when the input to the pipelined ADC is increasing, the control logic block makes the signal generator ramp down when the ADC output is A_j. Hence to make sure that the code A_{j+1} appears, the slope of the ADC should be increased. But the slope of the ramp should not be so high that a code higher than A_{j+1} appears at the ADC output. This will give erroneous test results. Similarly the slope of the ADC should not be so high that while ramping down, codes less than A_{j-1} appear. However, this increases the test time.

Considering all the above factors an optimum slope of the ramp has been selected. The ramp generator output is simulated by applying a square wave at the generator input.

5.3.4 Measurement of oscillation frequency

In order to evaluate the oscillation frequency coming from the circuit under test when the system is in test mode, it is to be measured by a digital counter, however, some signal conditioning is necessary before the signal is fed to the counter. Figure 5.17 shows a block diagram of the measurement technique for frequency. It uses a simple and fully digital circuit which converts each frequency to a related number. The oscillation frequency f_{osc} of the selected block is first passed through a Schmitt trigger (ST) to obtain a rectangular waveform compatible with logic levels and is then applied to a counter. The counter is enabled by the high level of enable voltage when the counter counts, and during its low state the counter is disabled, stops counting and holds the previous count. The output value of the counter contains a number which is related to its input frequency, coming from the block of

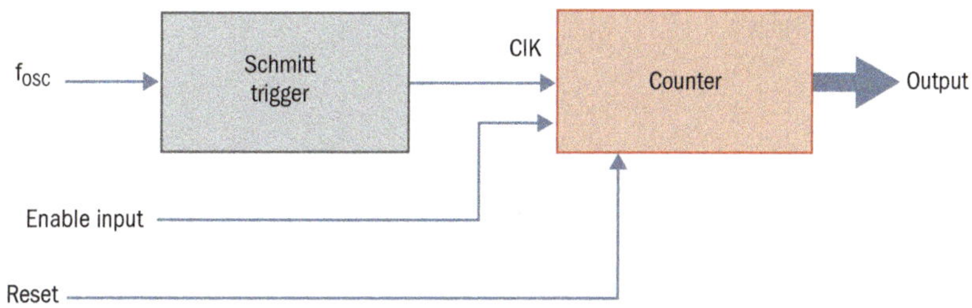

Figure 5.17. Block diagram of the FNC.

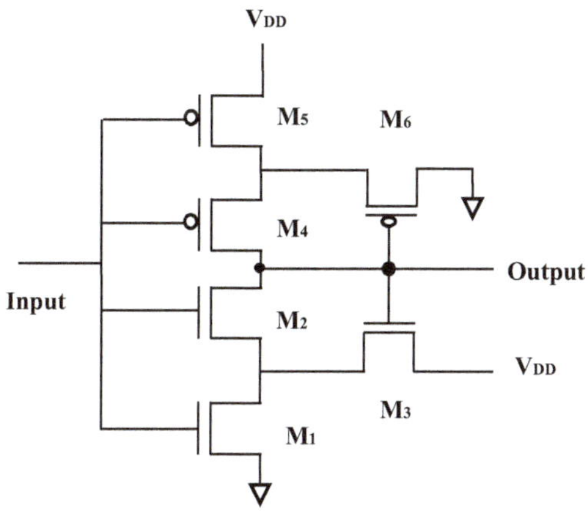

Figure 5.18. A CMOS Schmitt trigger.

the circuit under test, and can be evaluated by the control logic (CL) during the low state of enable input. A reset signal is necessary to make the counter ready for the next count. Thus, an accurate frequency-to-number conversion is obtained.

5.3.5 Schmitt trigger

A CMOS Schmitt trigger circuit is shown in figure 5.18. The hysteresis present in the transfer characteristics of the Schmitt trigger makes it different from the CMOS inverter. The input and output of a Schmitt trigger is shown in figure 5.19. As is evident from the illustration, it can convert any arbitrary shaped signal to a square wave that can be fed to the counter for its frequency measurement. When the output is high and the input exceeds the upper trip point, the output switches to low. When the input goes below the lower trip point the output will switch back to high.

5.3.6 Control logic block

The control logic block controls the entire BIST system and hence it is the most important part.

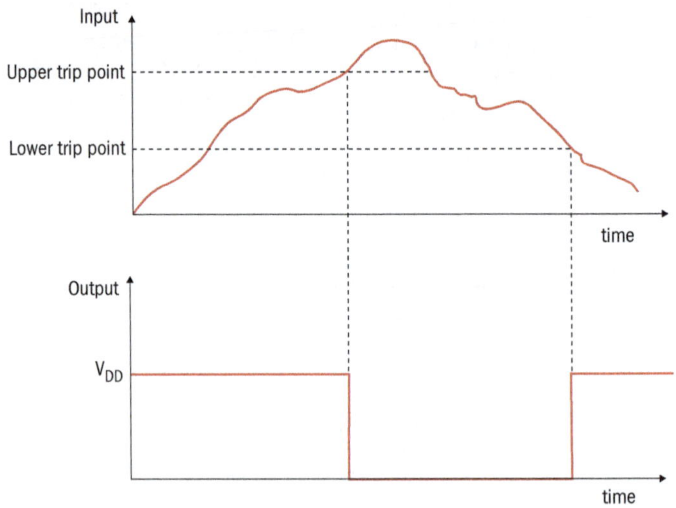

Figure 5.19. Output of a Schmitt trigger for an arbitrary shaped input signal.

5.3.6.1 Sub-block for control signal generation

The control logic block performs the very basic functionality of the block, i.e. creation of the oscillation in the ADC and hence this is the crux of the control logic block. For a particular reference code C_j, this sub-block compares the ADC output code to the reference code (the output code around which the ADC oscillates) and makes the control signal flag high when the ADC output is equal to or more than the reference code. This is achieved by a digital 8 bit subtractor unit, where the reference code is subtracted from the ADC output code. The two's complement of the reference code is taken and added to the ADC output code. When the ADC output code is lower than the reference code, no carry is generated. But when the ADC output is equal to or higher than the reference code the carry is 'HIGH'. This carry is used as the control signal [25]. The control logic block schematic is shown in figure 5.20.

5.3.6.2 Sub-block for reference code generation

For testing the DNL and INL errors for all the possible output codes of the pipelined ADC, the ADC should be made to oscillate around each code and then the linearity errors are measured during the ramp down period of oscillation. The reference generation sub-block generates the reference codes (around which the ADC has to oscillate and linearity errors are to be measured) automatically and sequentially from (0000…1) to (1111…1). To ensure least test time, after the linearity measurement for a particular code, the signal generator is not reset and again increased from zero. For a particular reference code C_j, when the ADC output produces C_{j-1} after a complete oscillation around code C_j (C_{j-1}–C_j–C_{j+1}–C_j–C_{j-1}), the reference code generator code is increased by 1 LSB to generate the next reference code and make the signal generator ramp up for the next oscillation. Hence for a particular reference code C_j, the ADC is made to oscillate around it only once

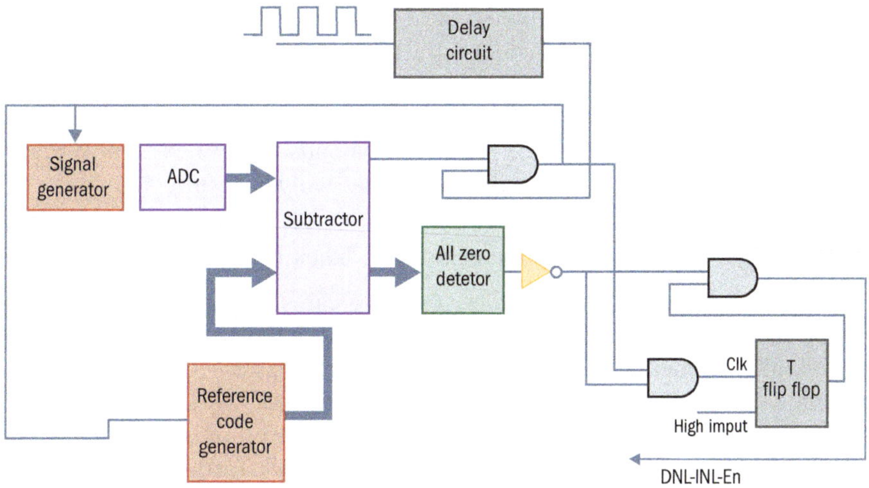

Figure 5.20. Schematic of the control logic unit.

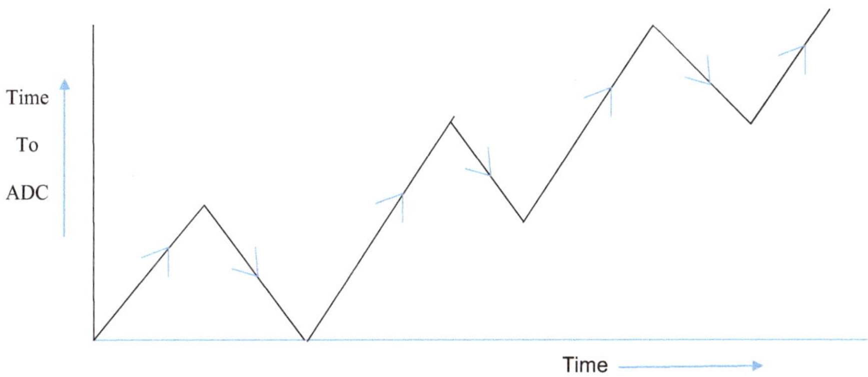

Figure 5.21. Output characteristics of the signal generator.

(to save test time) and after the first oscillation is complete, the reference code is changed to the next higher code and the ADC input ramps up for the next oscillation [26].

This reference code generation is performed by an up-counter which is initialized with (0000…1). During the initial delay period the D flip-flop in the reference generation circuit is disabled. Only after the first ramp down mode has been started is the D flip-flop activated and its output becomes 'HIGH'. But only when the control signal again becomes low is the counter increased by 1 LSB. The input to the ADC follows the profile shown in figure 5.21.

5.3.7 DNL and INL measurement blocks

In the DNL measurement block the output of the FNC is sent to a 8 bit subtractor in which the FNC's output is subtracted with an ideal value of 8 bits corresponding to an ideal ADC, then it is divided by the ideal value count to normalize it with respect to the LSB (figure 5.22). To achieve this, first the output of the subtractor is multiplied by decimal value 100 (bit value 1100100) through a 7 × 7 multiplier and then the product is sent to a 15 by 8 bit divider circuit in which the product is divided by the ideal 8 bit count value. The result corresponds to the DNL value multiplied by 100. So if the DNL is 0.48 the divider output will show only 48.

5.3.8 Reconfigurability for detection of a faulty block in the ADC

Many times after post-fabrication there occurs a catastrophic fault, which is due to the fabrication error or mask error and in which 1 or 2 individual blocks of the ADC become faulty. Under this circumstance it becomes extremely difficult for the IC chip designer to diagnose exactly which blocks are not working. Keeping this in mind, the reconfigurability feature of the OBIST architecture is being introduced, in which the AMUX selects the ADC in a block-by-block form and to each block a triangular wave is sent as the input around the LSB value of the ADC, and the output is sent to the FNC (figure 5.23). The blocks which show a greatly deviated result from the average value can be considered as faulty and thus can be corrected in the pre-fabrication stage [26].

5.4 Implementation of the ADC and OBIST

5.4.1 The pipelined ADC

To check whether our BIST architecture will work or not, or if it will work then how much better results it can produce compared to the traditional testing schemes, we needed a pipelined ADC with moderate specifications. Thus we designed a pipelined ADC in the Cadence Virtuoso environment (180 nm technology). The specifications are as shown in table 5.1.

5.4.2 Design implementation of the blocks of the pipelined ADC

In this section the design of all the sub-blocks will be discussed. The circuit schematic and the simulation results are also presented here.

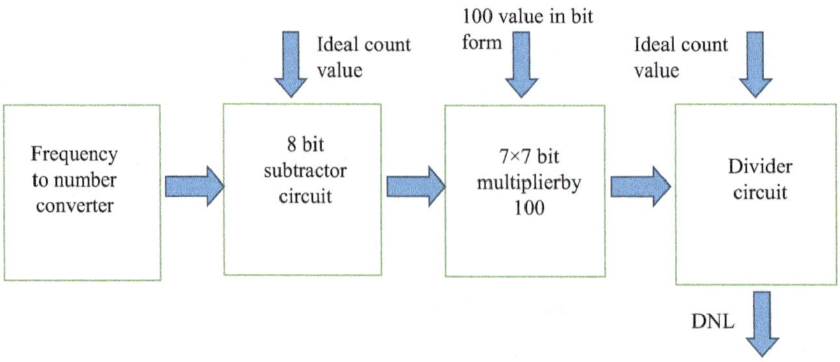

Figure 5.22. Flow diagram of the DNL measurement block.

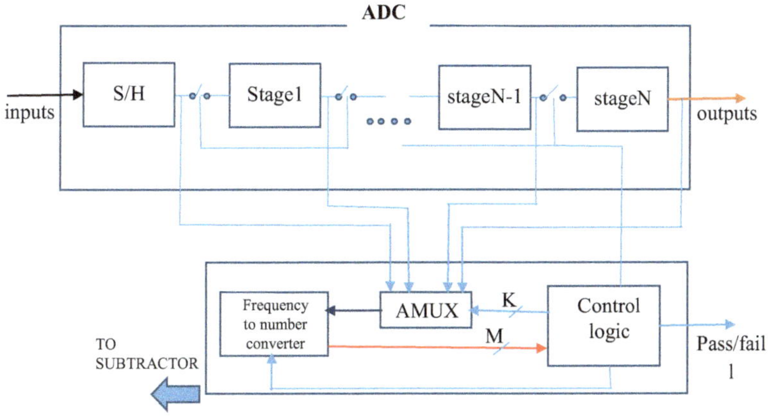

Figure 5.23. Architecture for the reconfigurability scheme.

Table 5.1. Specifications of the pipelined ADC.

Specifications	Value
Size	8 bit
Gain (OTA)	54 dB
Unity gain bandwidth	106 MHz
ICMR	300 mV
Slew rate	190 V μs^{-1}
Sampling rate	40 MHz

5.4.2.1 Operational amplifier

We designed a folded cascode amplifier with a gain of 54 dB, a unity gain bandwidth of 106 MHz and a phase margin of 80 degrees. which allows for some losses due to the parasitic components added during the layout. The folded cascode amplifier and the circuit schematic are shown in figures 5.24 and 5.25, respectively. The AC analysis of the designed circuit is shown in figure 5.26. It is found that the circuit simulation results will be on par with the specifications laid out at the beginning of this section [17].

5.4.2.2 S&H circuit

After the amplifier the S&H circuit was designed as it is the first stage of the ADC. The circuit schematic of the S&H circuit is shown in figure 5.27. The transient analysis of the S&H circuit is shown in figure 5.28.

5.4.2.3 Comparator

The comparator used in this work is a differential dynamic comparator and is shown in figure 5.29 [27]. Because of the dynamic current sources, together with the latch, connected directly between the differential pairs and the supply voltage, the

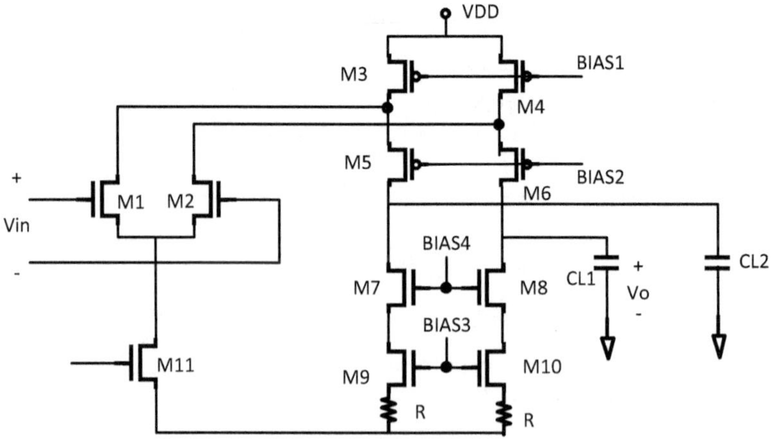

Figure 5.24. Design of the folded cascode amplifier.

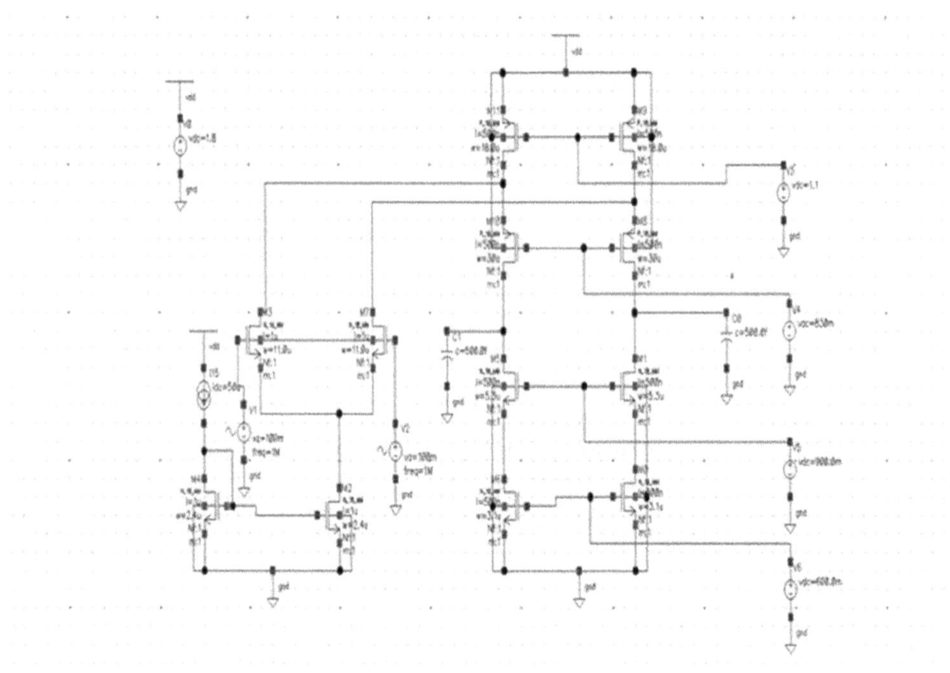

Figure 5.25. Schematic of the folded cascode amplifier.

comparator does not dissipate dc power. When V_{latch} is raised to V_{DD}, the outputs are disconnected from the positive supply, and the switching current source turns on and $(V_{\text{in}+} - V_{\text{in}}-)$ is compared with $(V_{\text{ref}+} - V_{\text{ref}-})$. The circuit schematic of the differential dynamic comparator and its transient analysis are shown in figures 5.30 and 5.31, respectively.

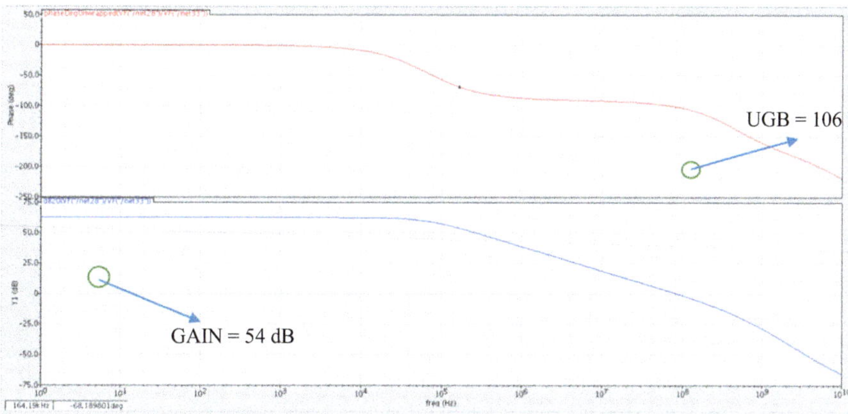

Figure 5.26. AC analysis of the folded cascode amplifier.

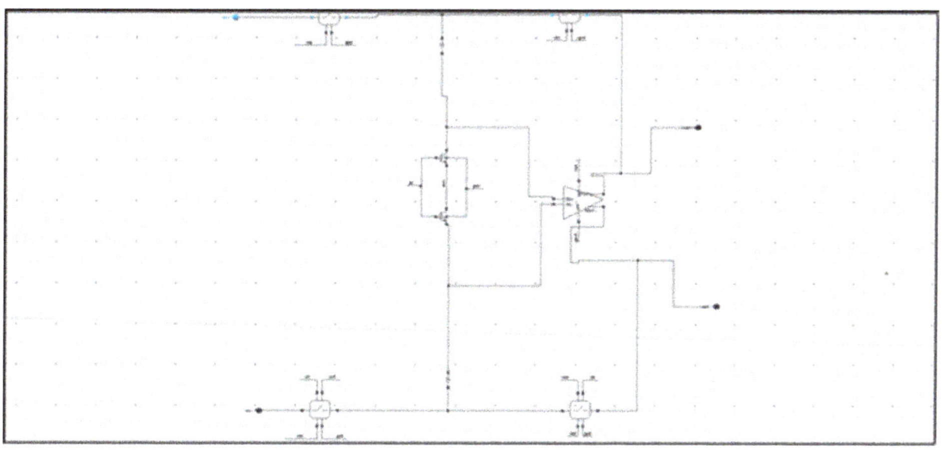

Figure 5.27. Schematic of the S&H circuit.

5.4.2.4 Sub-ADC and DAC

The quantization is performed by the sub-ADC block which resolves 1.5 bits of resolution and provides the necessary values for subtraction from the input value which, after being amplified by 2, results in the residual signal for the next stage. Each sub-ADC contains two comparators and one DAC. In the DAC section the outputs of two comparators are evaluated to give a single output which is listed as follows [16]. The circuit schematic of sub-ADC is shown in figure 5.32.

- $V_{ref+} - V_{ref-}$ when $V_{in} > V_{ref}/4$.
- V_{CM} when $-V_{ref}/4 < V_{in} < V_{ref}/4$.
- $V_{ref-} - V_{ref+}$ when $V_{in} < V_{ref}/4$.

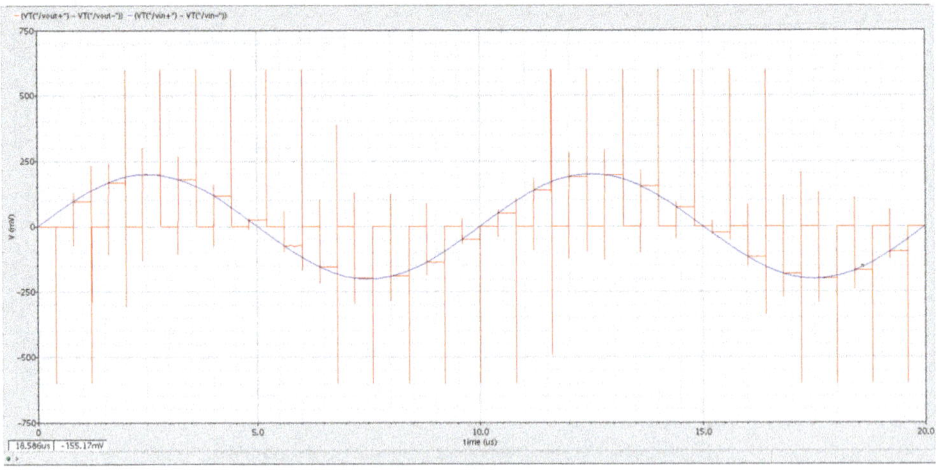

Figure 5.28. Transient analysis of the S&H circuit.

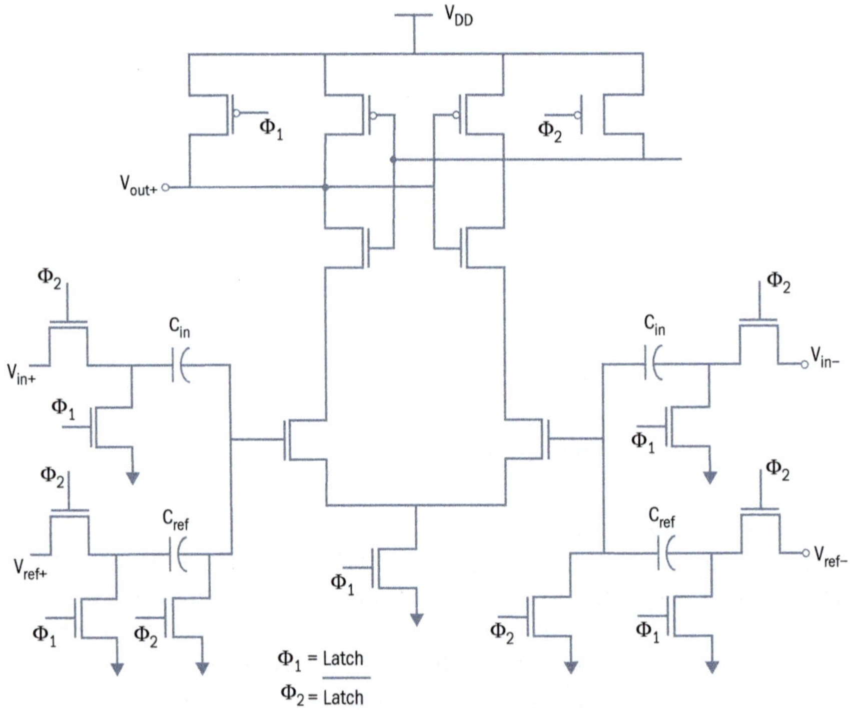

Figure 5.29. Differential dynamic comparator.

5.4.2.5 The pipelined stage

Figure 5.33 shows the typical pipelined stage in Cadence. It consists of a sub-ADC and switched-capacitor amplifier. The simulation result of one stage with sine wave input is shown in figure 5.34.

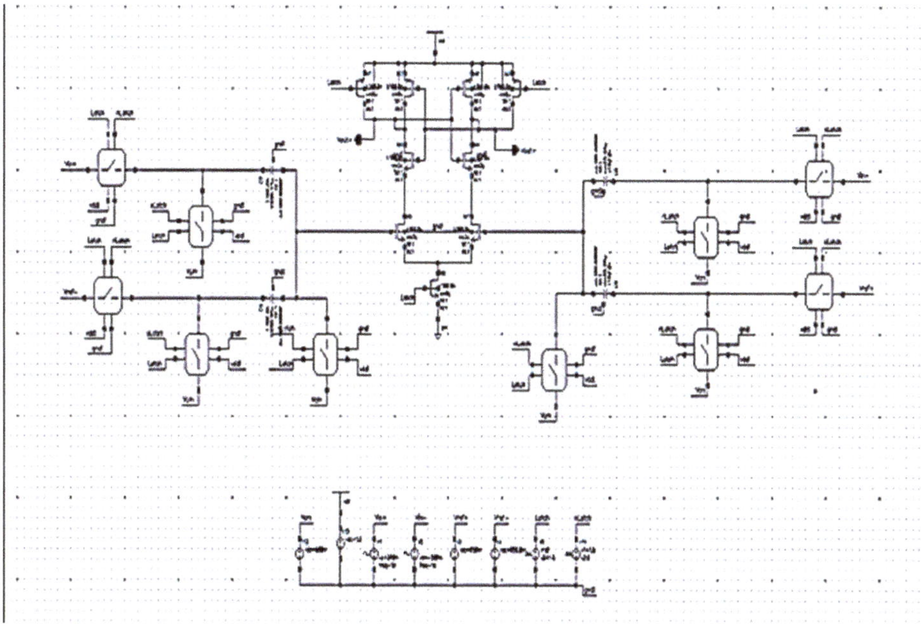

Figure 5.30. Schematic of the differential dynamic comparator.

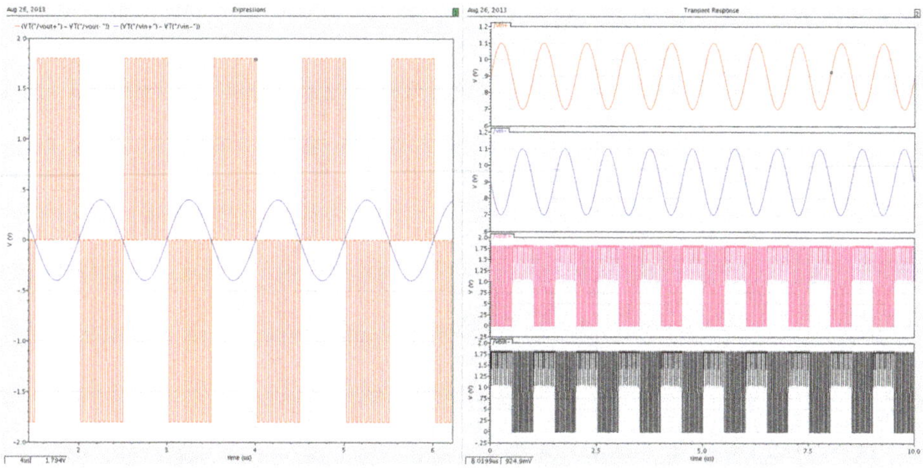

Figure 5.31. Transient analysis of the differential dynamic comparator.

5.4.2.6 The last stage

In order to reduce the area and power consumption of the ADC, the design of the last stage is considered. By using a 2 bit converter instead of a simple comparator for the last stage we could remove one full pipelined stage from our design as depicted in figure 5.35. The transient analysis is shown in figure 5.36.

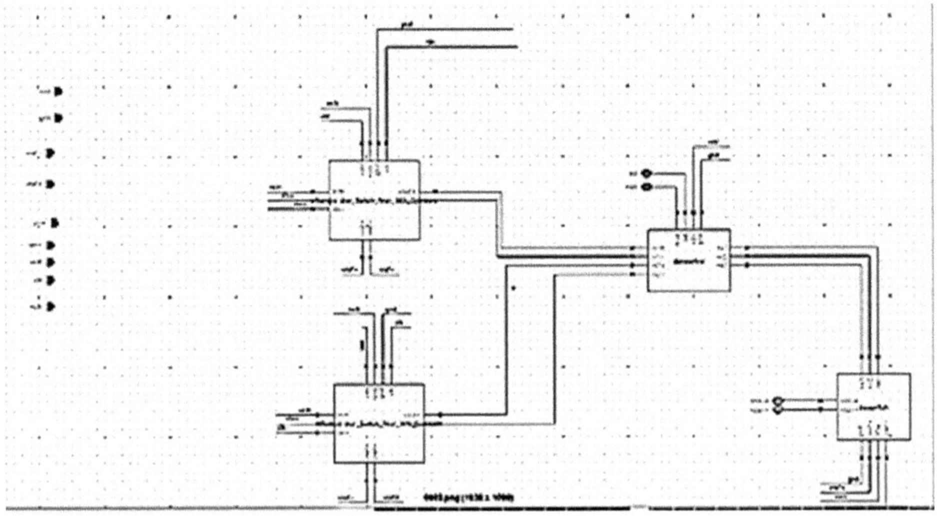

Figure 5.32. Schematic of the sub-ADC in Cadence.

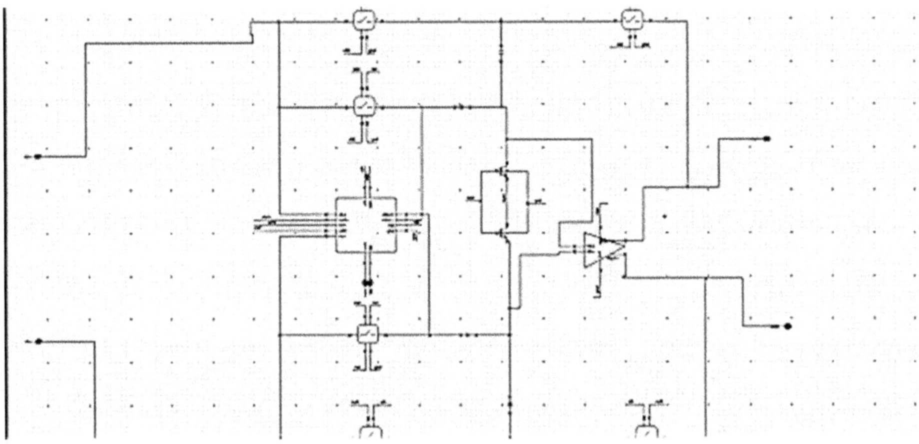

Figure 5.33. Schematic of the pipelined stage in Cadence.

5.4.2.7 Complete ADC

Finally, we put all stages together to make the pipelined ADC. It is shown in figure 5.37. The first block in the illustration, as mentioned above, is the S&H block—it is continued with six typical stages and finished with the last stage. The blocks on the right-hand show delay and digital correction. Many simulation results are taken and also the output of the ADC is sent to the DAC so that the input can be shaped back. The transient analysis of the complete pipelined ADC is shown in figure 5.38. The transient analysis of the ADC followed by the DAC is shown in figure 5.39.

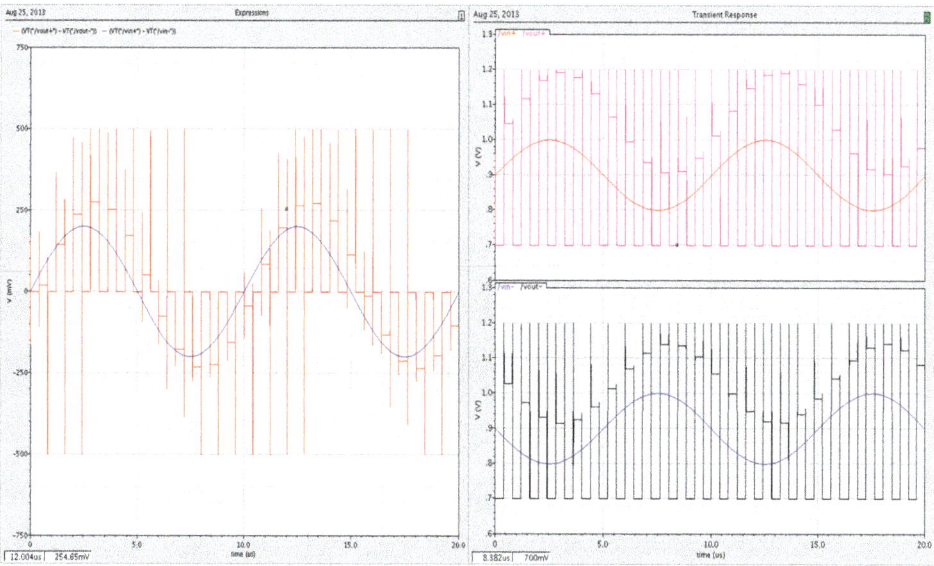

Figure 5.34. Transient analysis of the pipelined stage.

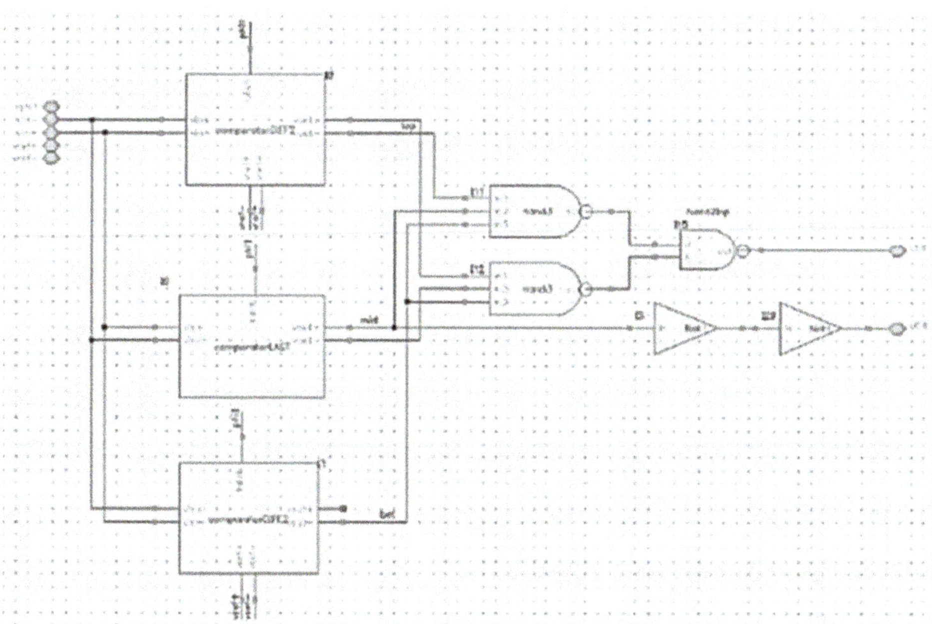

Figure 5.35. Schematic of the last stage of the pipelined ADC.

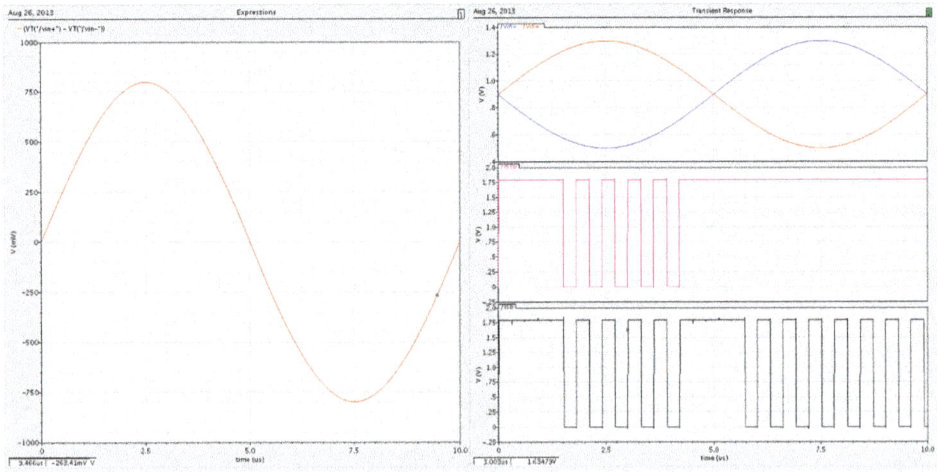

Figure 5.36. Transient analysis of the last stage of the pipelined ADC.

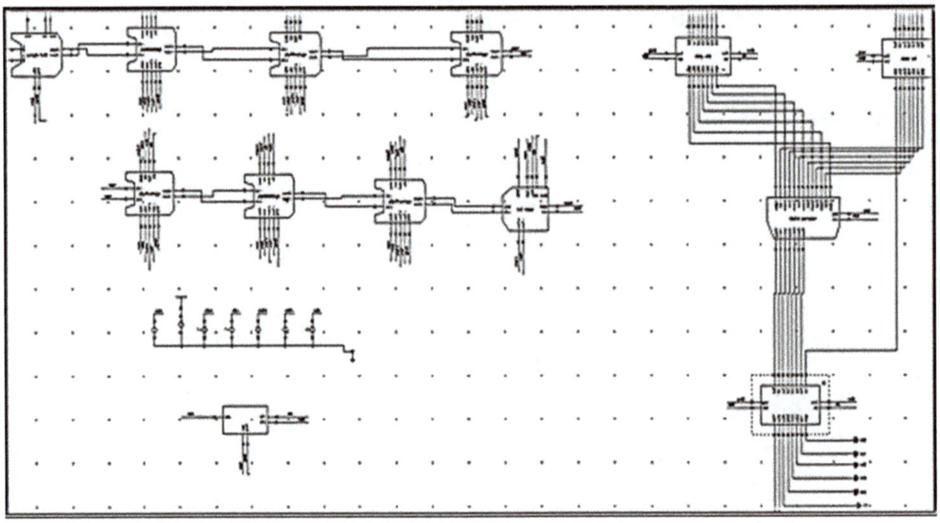

Figure 5.37. Schematic of the complete pipelined ADC in Cadence.

5.4.3 Design implementation of OBIST

5.4.3.1 Level crossing detector

The level crossing detector is part of the frequency-to-number conversion block followed by an 8 bit counter. The level crossing detector changes the oscillatory output of the ADC, converted it into a square wave from which frequency can be deduced easily by the counter. The transient analysis is shown in figure 5.40.

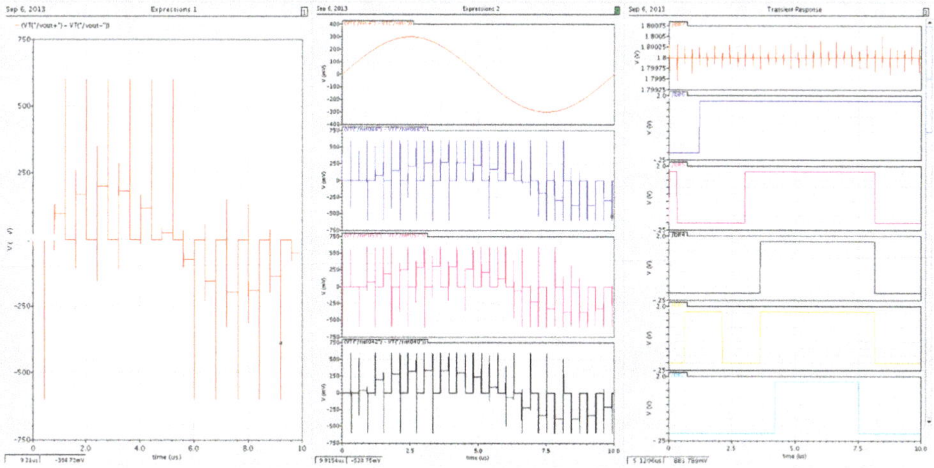

Figure 5.38. Transient analysis of the complete pipelined ADC.

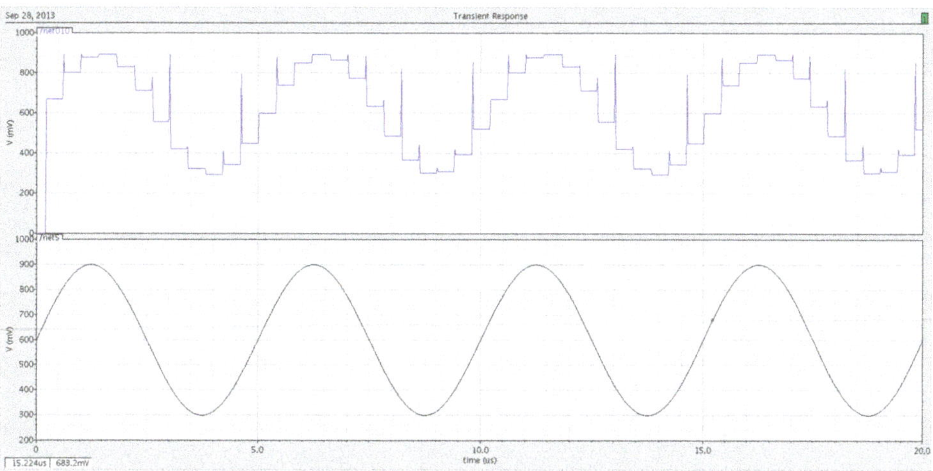

Figure 5.39. Transient analysis of the ADC followed by the DAC.

5.4.3.2 Counter

We designed an 8 bit counter using *J–K* flip-flop although 6 bits can also achieve the purpose. The counter will count up to the up level of f_{ref} and after that the counter will be reset. A healthy ADC will always give a signature count and a faulty one will give a different count. The counter schematic and its transient analysis are shown in figures 5.41 and 5.42, respectively.

5.4.3.3 Current reference design

This is the most important part of the OBIST architecture. This circuit produces a triangular wave around the threshold transition voltage of the ADC for a specific code when a switching frequency of value f_{osc} is given to the circuit. The circuit

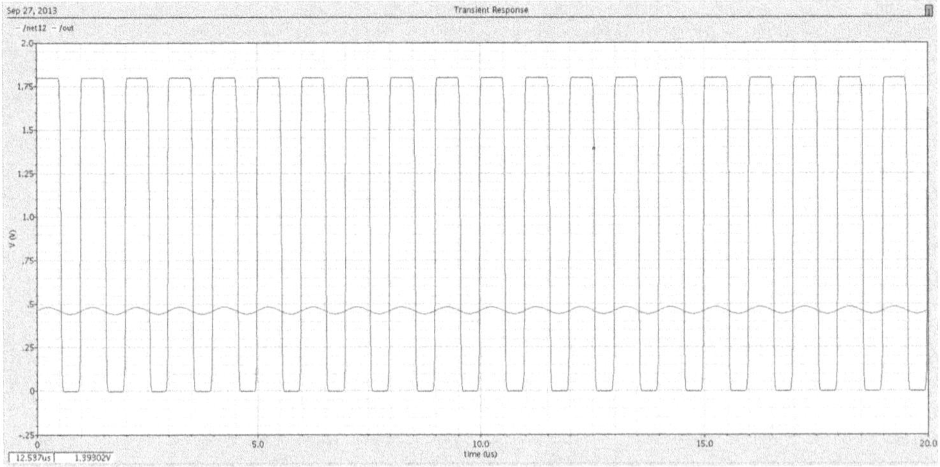

Figure 5.40. Transient analysis of the level crossing detector.

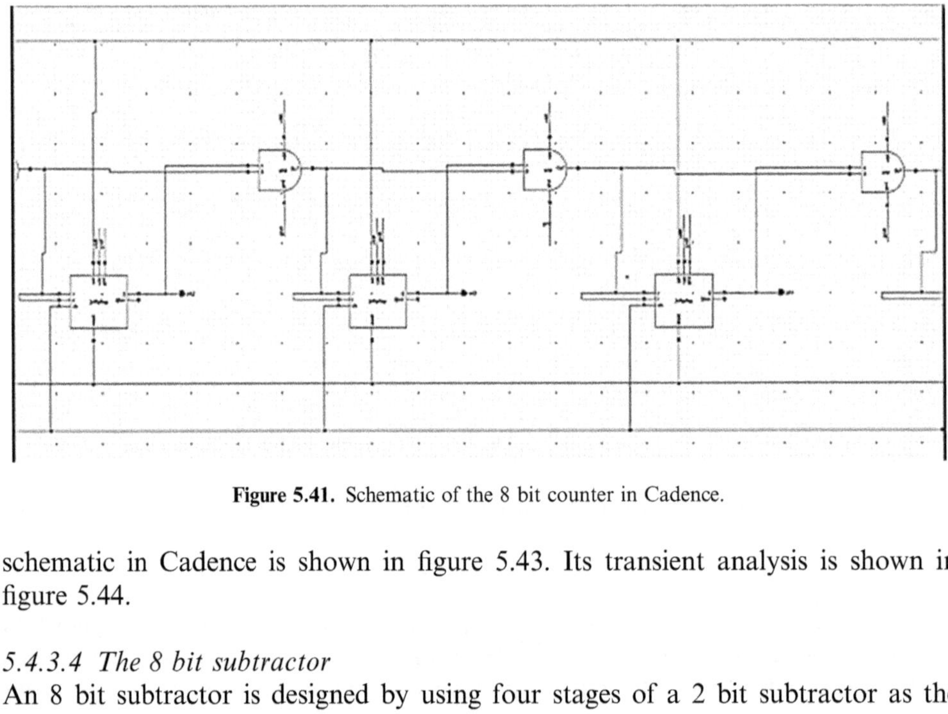

Figure 5.41. Schematic of the 8 bit counter in Cadence.

schematic in Cadence is shown in figure 5.43. Its transient analysis is shown in figure 5.44.

5.4.3.4 The 8 bit subtractor

An 8 bit subtractor is designed by using four stages of a 2 bit subtractor as the building blocks. The schematics of the basic 2 bit and 8 bit subtractors are shown in figures 5.45 and 5.46, respectively.

5.4.3.5 Multiplier

A 7×7 multiplier is used to multiply the output of the 8 bit subtractor by 100 (bit value 1100100) because if we do not multiply the result by 100 we cannot divide directly by the ideal counter count because the DNL value will be in decimals (<1)

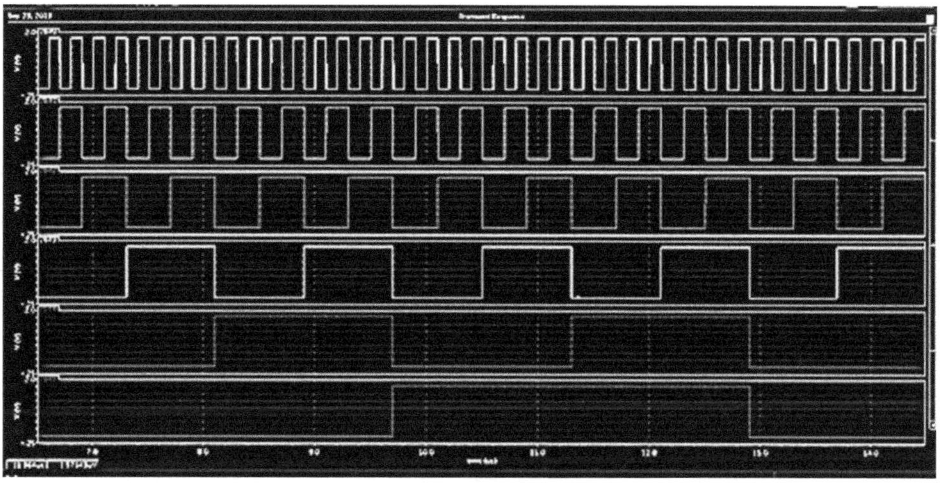

Figure 5.42. Transient analysis of the 8 bit counter.

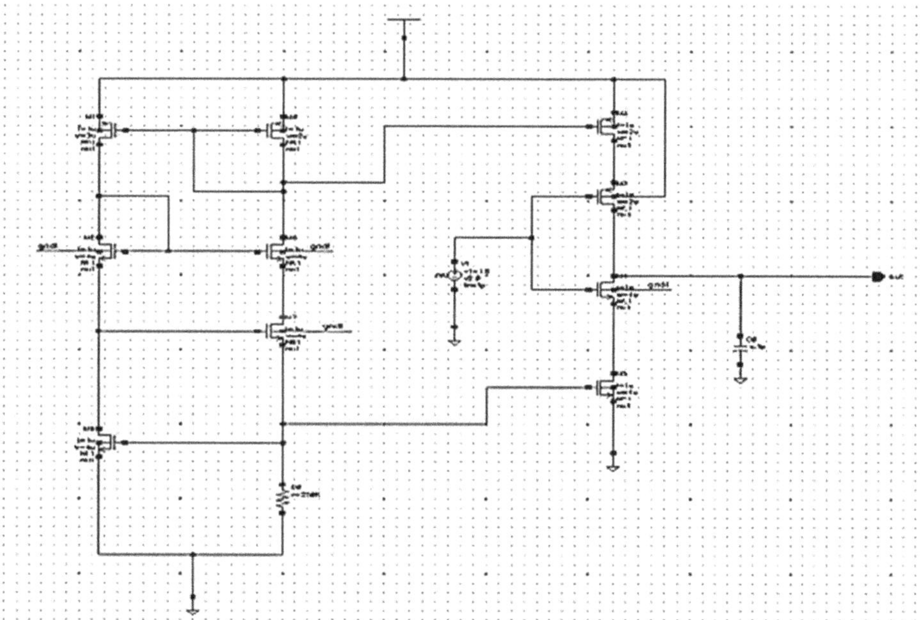

Figure 5.43. The circuit schematic.

and we cannot perform the division in bit format for the decimal part. First a basic building block was designed (the multiplying full adder (MADD)) and the block was replicated to construct a 7 × 7 multiplier circuit [16]. The architecture of the multiplier block is shown in figure 5.47. The schematics of the MADD block and multiplier circuit in Cadence are shown in figures 5.48 and 5.49, respectively.

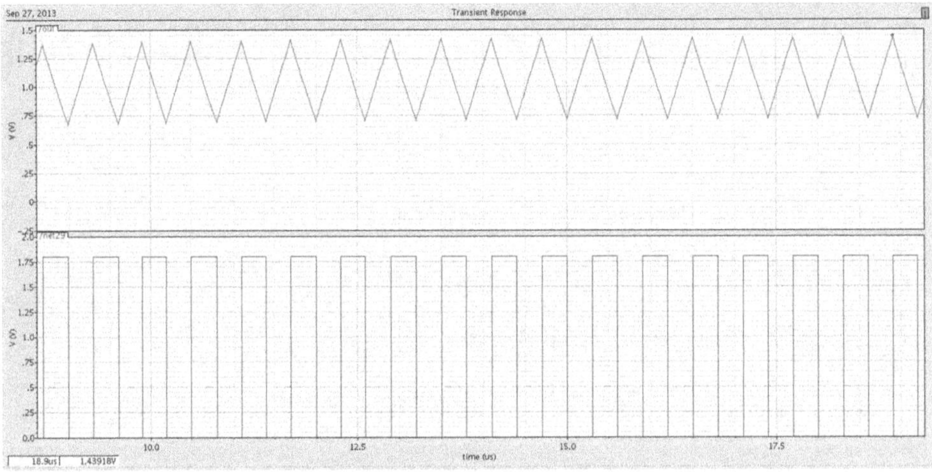

Figure 5.44. Transient analysis of the current reference generator.

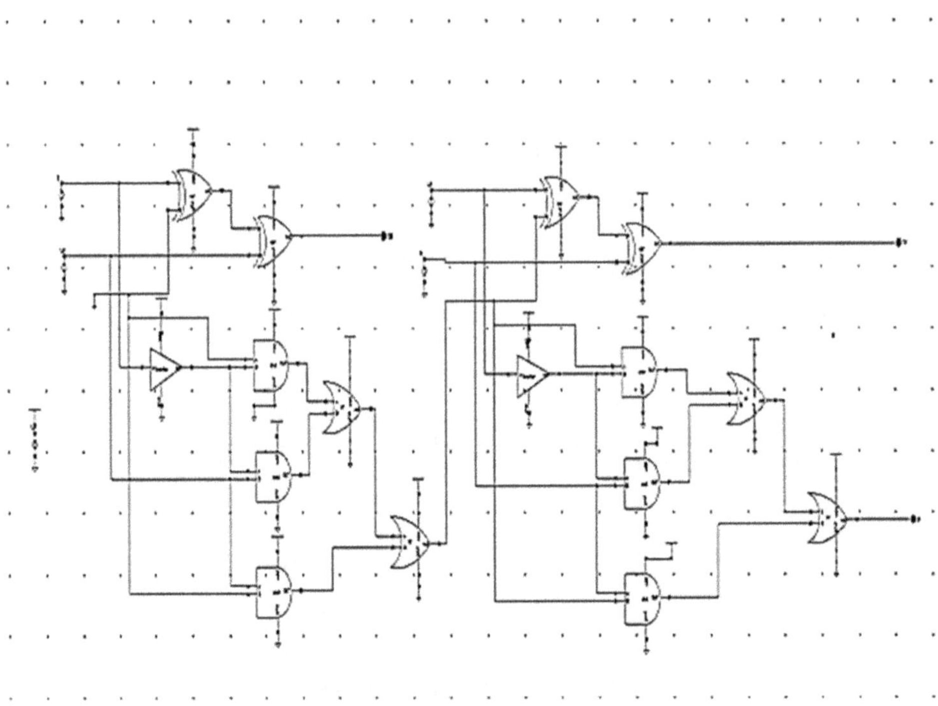

Figure 5.45. Schematic of a 2 bit subtractor circuit in Cadence.

5.4.3.6 A 16 by 8 bit divider block

A divider circuit is designed to divide the 14 bit product from the multiplier block by the 8 bit ideal count value to give a 8 bit DNL value as the quotient (figures 5.50 and 5.51). A basic controlled subtractor and adder (CSA) block was designed first and by replicating it the whole circuit was constructed [20].

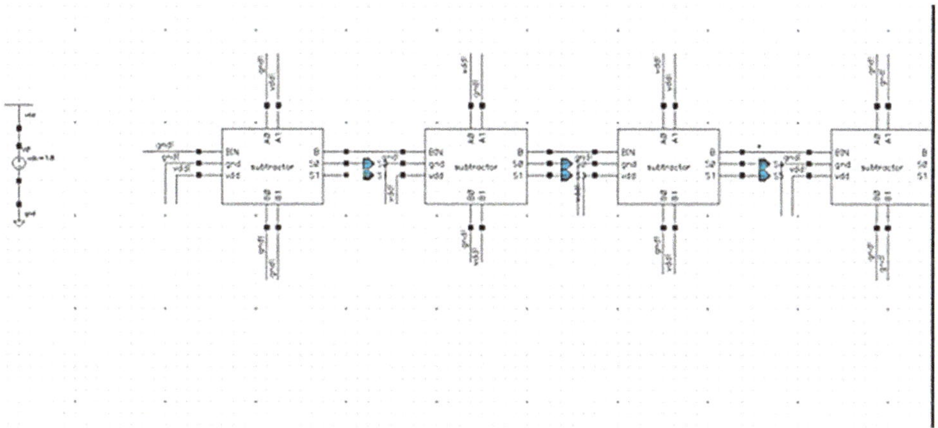

Figure 5.46. Schematic of an 8 bit subtractor circuit in Cadence.

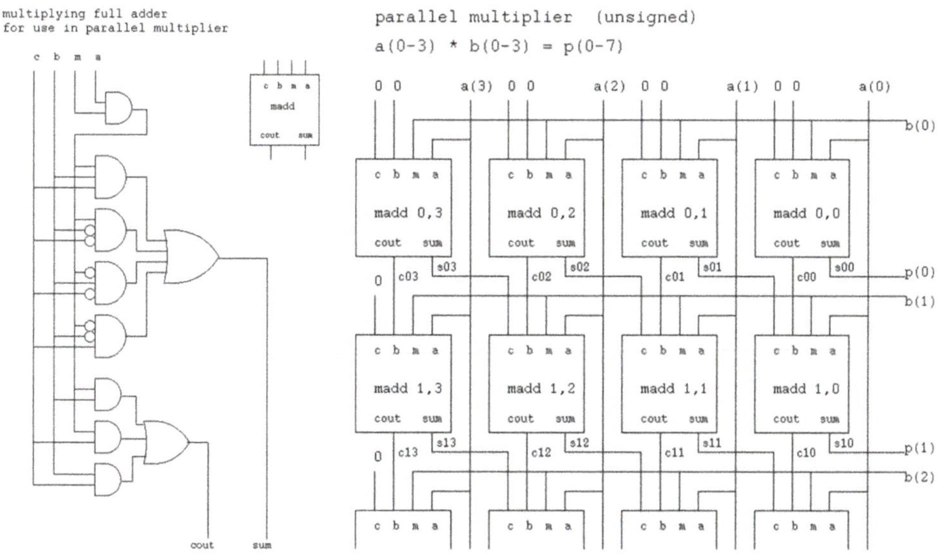

Figure 5.47. Architecture of the multiplier block.

5.4.4 Test bench

A test bench to check the validity of the architecture is created and the results are also evaluated for the DNL error. An ideal 8 bit ADC designed through Verilog A was used as standard to calculate the DNL and INL. The count value from an ideal ADC is used in the subtractor and divider blocks to calculate the DNL error (figures 5.52–5.56).

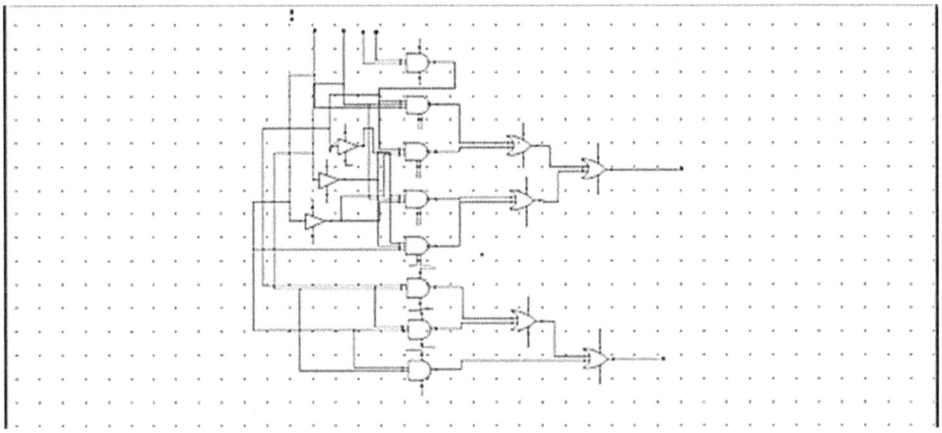

Figure 5.48. Schematic of the MADD block in Cadence.

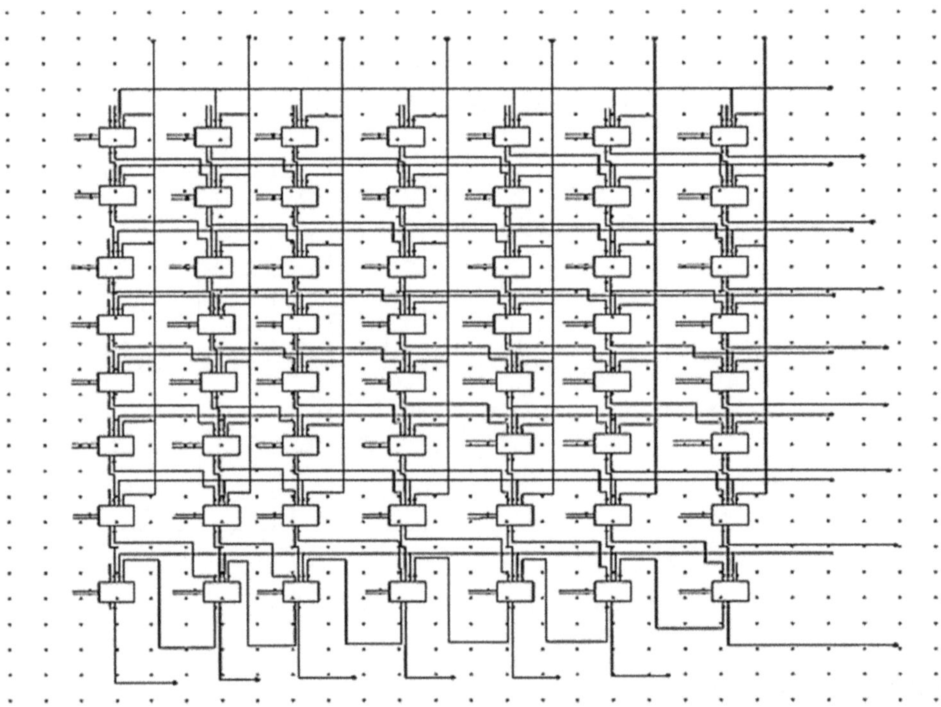

Figure 5.49. Schematic of a 7 × 7 multiplier circuit in Cadence.

5.4.5 Simulation results

The architecture is tested for an ideal ADC and our pipelined ADC simultaneously and the counter value is checked at a particular code transition (figures 5.57 and 5.58). As was expected, the values were different. We calculate the DNL as

$$\text{DNL} = (\text{Counter Value}_{\text{practical}} - \text{Counter Value}_{\text{ideal}})/\text{Counter Value}_{\text{ideal}}.$$

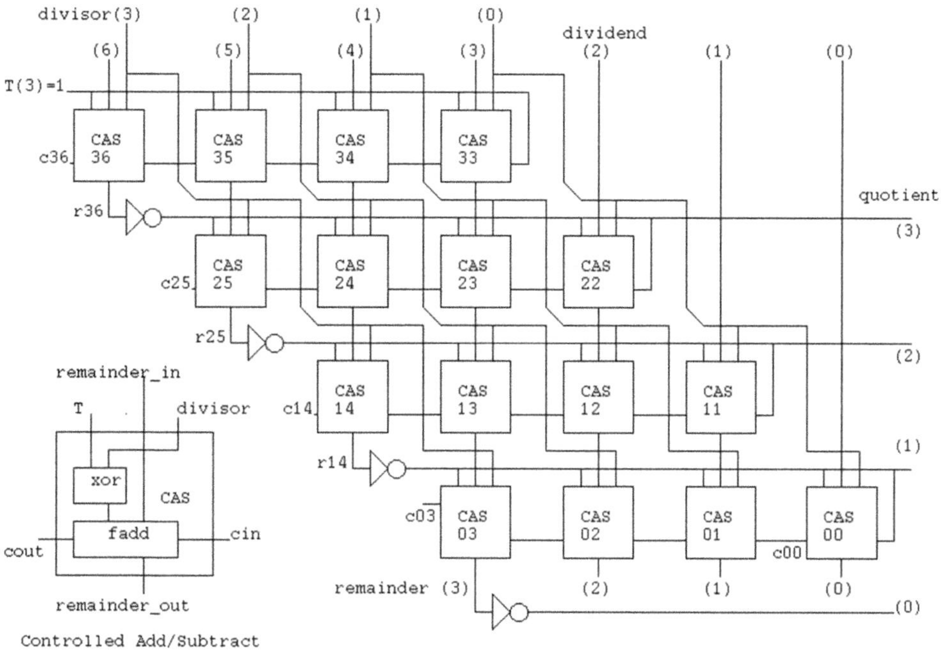

Figure 5.50. Architecture of an 8 by 4 bit divider circuit.

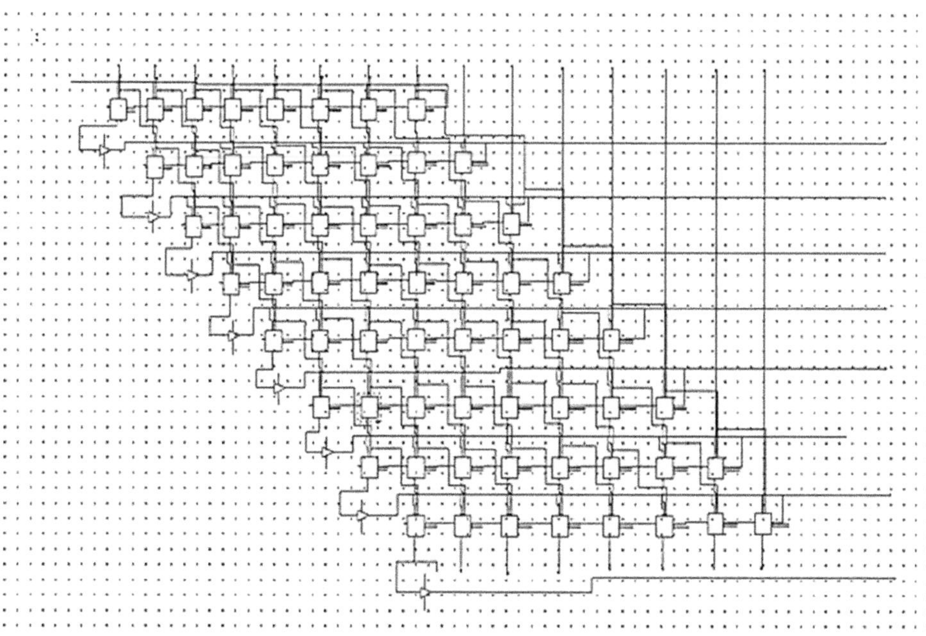

Figure 5.51. Schematic of an 16 by 8 bit divider circuit.

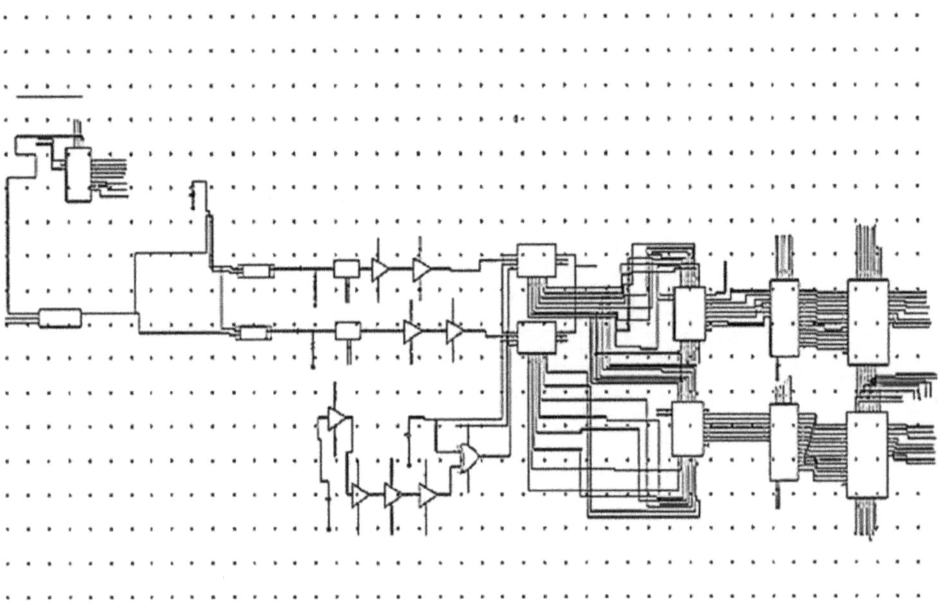

Figure 5.52. Schematic of the test bench set-up in Cadence.

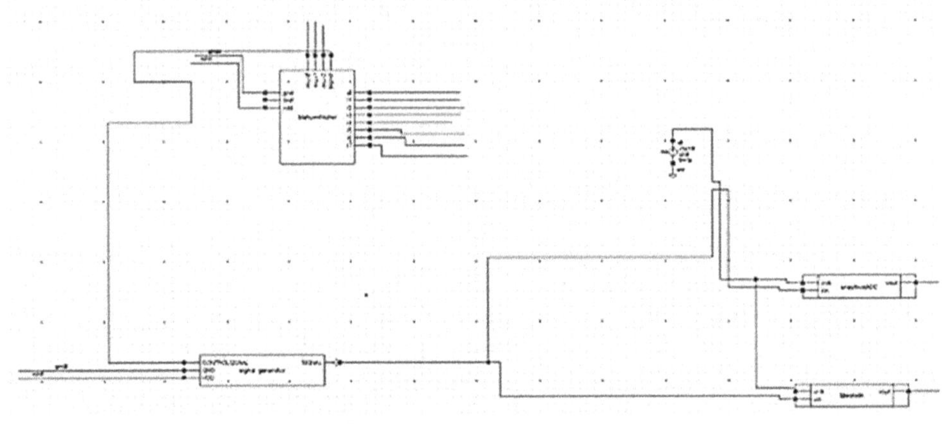

Figure 5.53. Schematic of the first part of the test bench from a left side view.

Theoretically DNL was found to be (110110 binary − 101000 binary)/ 101000 = 0.35 LSB, which is less than the 1 LSB that indicates the validity of the pipelined ADC as well as of the BIST architecture.

Figure 5.59 shows the result of 110110 binary − 101000 binary[Counter $\text{Value}_{\text{practical}}$ − Counter $\text{Value}_{\text{ideal}}$] which is equal to 001110 binary.

The output of the multiplier showing the result of 001110 × 1100100 (decimal value 100) = 10101111000 is shown in figure 5.60.

Figure 5.61 shows that the output of the divider blocks is 00100011 binary, which is equivalent to 35 in decimal. So our DNL error comes out as 0.35 LSB.

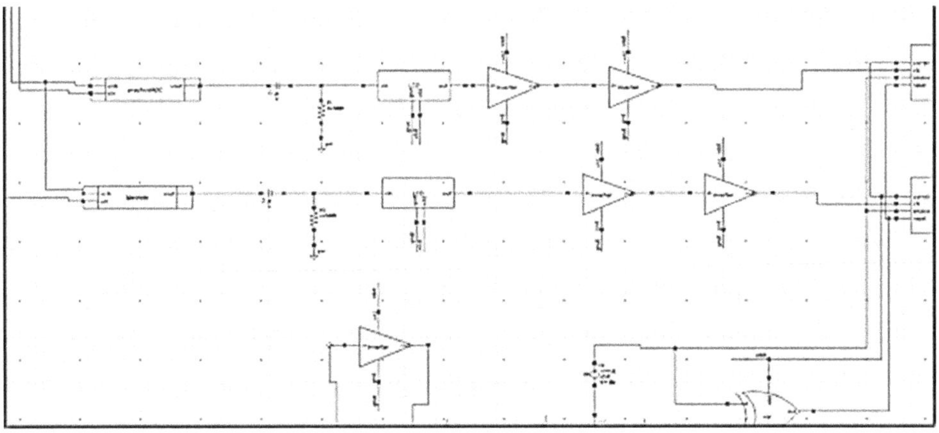

Figure 5.54. Schematic of the second part of the test bench from a left side view.

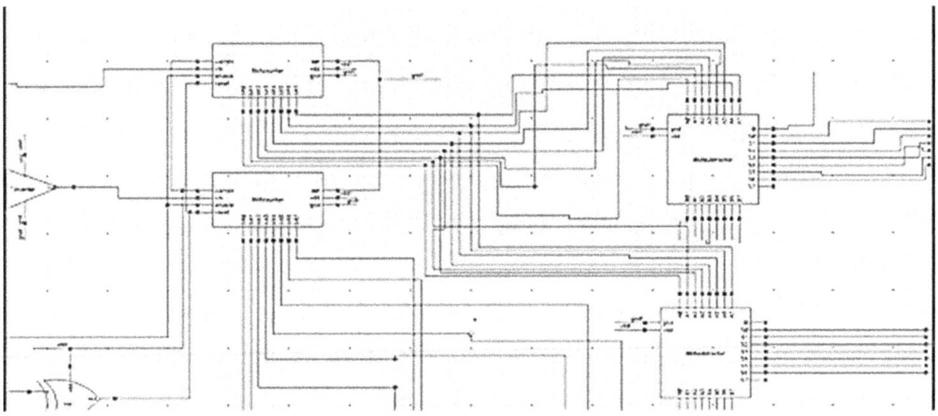

Figure 5.55. Schematic of the third part of the test bench from a left side view.

5.4.6 Diagnosis of a faulty block by reconfigurability

For the diagnosis of a faulty block by reconfiguring the OBIST architecture, we first design an analog multiplexer (AMUX) and then demultiplexer (DEMUX) to isolate a particular block of the ADC (figures 5.62 and 5.63). Then with the help of a signal generator we produce a triangular wave around the transition voltage level for 1 LSB of the ADC (1/256 of 1.8 V). Then we observe the counter value of each of the blocks for different faults intentionally generated by us.

We can see from table 5.2 that the scheme can diagnose faults in every aspect. However, we have to keep in mind that for the faults corresponding to the DAC and op-amp we have to isolate two blocks at a time and the count value will be different in the second block of the configuration.

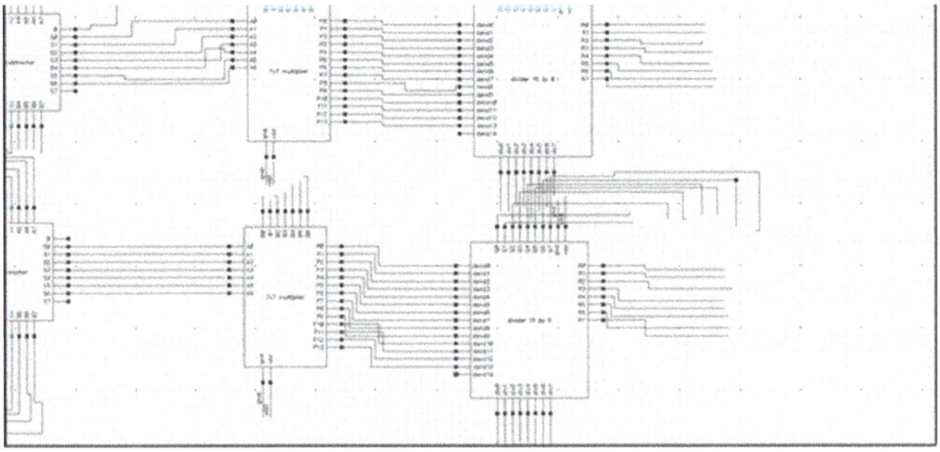

Figure 5.56. Schematic of the fourth part of the test bench from a left side view.

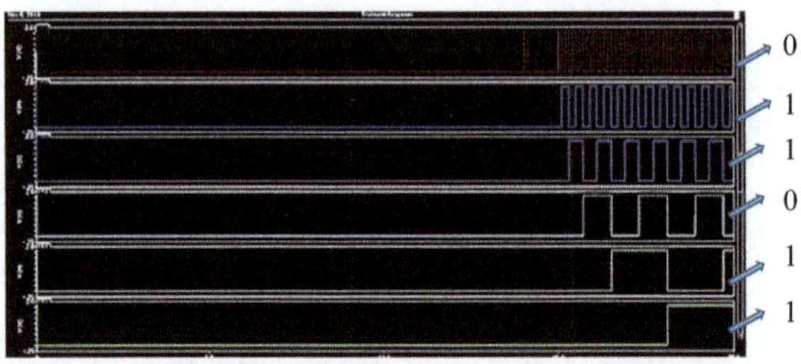

Figure 5.57. Counter value for a practical pipelined ADC.

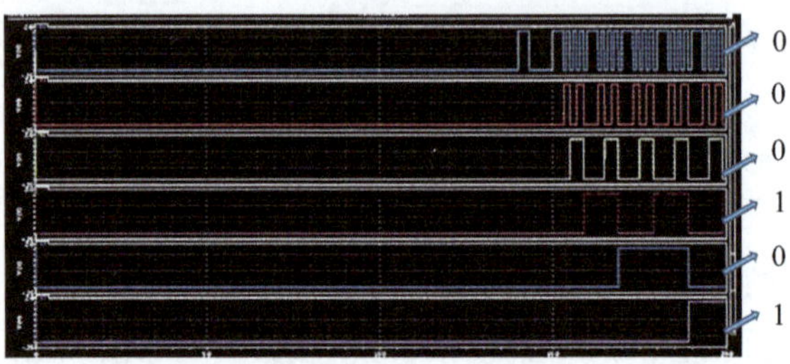

Figure 5.58. Counter value for an ideal ADC.

5.5 Conclusion

5.5.1 Work done

In this work a novel oscillation-based BIST system has been designed for a 1.8 V, 8 bit pipelined ADC in 180 nm standard CMOS technology in the Cadence environment. Oscillation-based BIST (OBIST) is a general BIST scheme applicable to analog and mixed-signal circuits. This technique has been previously applied on analog filters and operational amplifiers. In this work the test technique is applied for the functional testing of a pipelined ADC. The ADC is reconfigured to oscillate

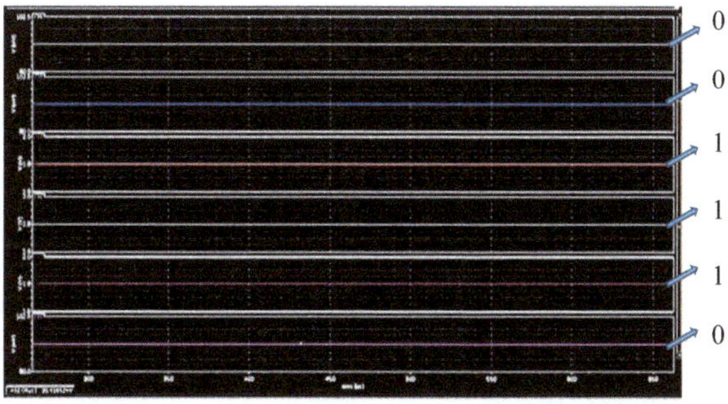

Figure 5.59. Output of the 8 bit subtractor circuit.

Figure 5.60. Output of the 7 × 7 multiplier circuit.

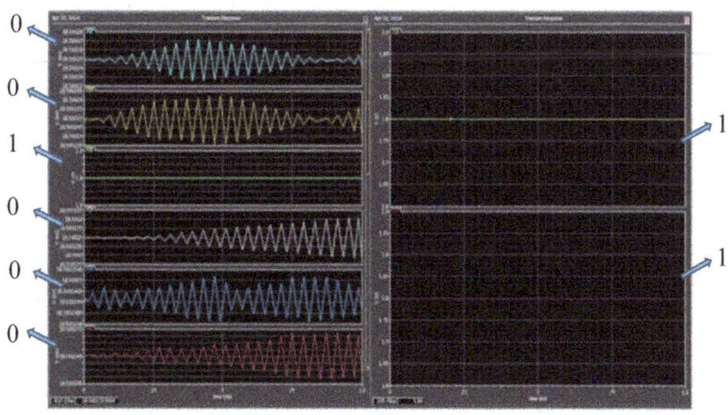

Figure 5.61. Output of the 16 by 8 divider circuit.

around a particular output code and from the code-width measurement the differential non-linearity (DNL) error is measured.

A 1.8 V, 8 bit pipelined ADC was built in the Cadence Virtuoso environment to test the OBIST methodology. The pipelined ADC was designed block by block and simulation results were shown for each component and block. Then the ADC was integrated and OBIST blocks were designed, namely the FNC current reference generator, 8 bit subtractor, 7 × 7 multiplier and 16 by 8 bit divider circuit.

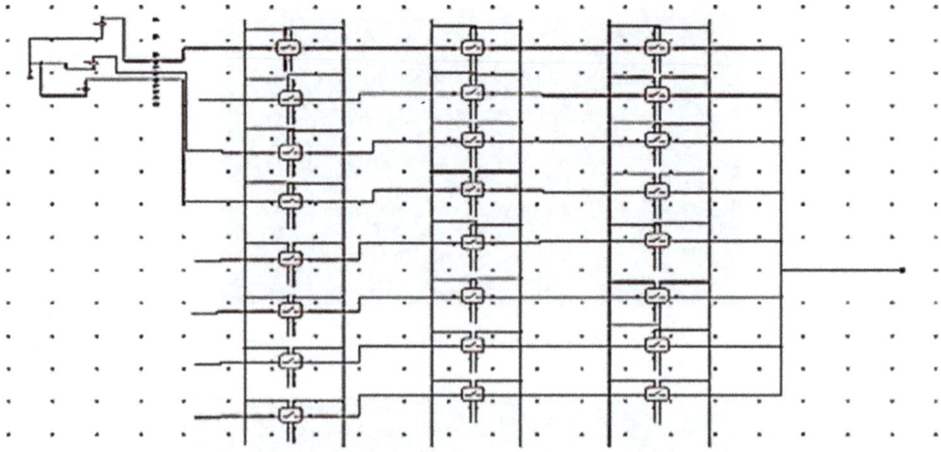

Figure 5.62. Schematic of the analog multiplexer in Cadence.

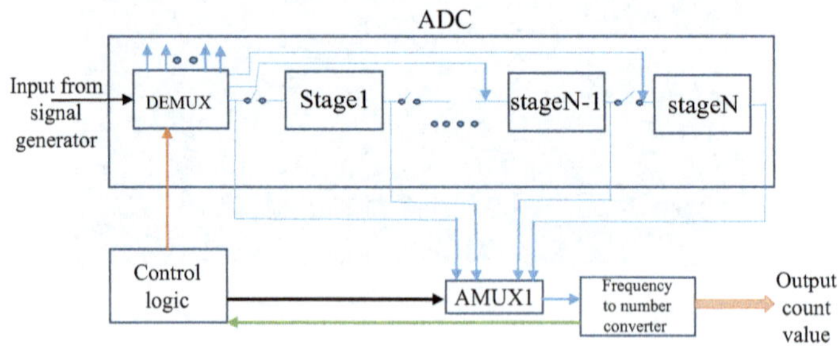

Figure 5.63. Architecture for a reconfigurability scheme.

Table 5.2. Counter values for different blocks in tabular form.

FAULTY REGION → / STAGE ↓	COMPARATOR	OP-AMP (2 blocks at a time)	SUB-ADC	DAC (2 blocks at a time)
STAGE 1	44	38	44	25
STAGE 2	44	38	44	25
STAGE 3	47	40	47	26
STAGE 4	45	38	45	25
STAGE 5 (FAULTY)	00	40	20	26
STAGE 6	45	12	45	09
STAGE 7	45	38	45	25

A test bench was designed and simulation was performed for finding the DNL error. After successfully evaluating DNL errors, we came to the diagnosis of a faulty block through AMUX. Different faults were created and simulation results were presented in tabular form. The results show that this scheme is capable of diagnosing the faulty block. Summarizing our work we can conclude that:

- A test technique for complete and accurate functional testing of ADCs has been developed using the proposed OBIST technique.
- The validity of the OBIST technique for ADCs has been verified through extensive simulation.
- The results show that the OBIST architecture is suitable for both functional and structural testing and has the capability of diagnosing faulty blocks of ADCs.
- As the OBIST structure presents a pass or fail result, it can be easily integrated with the test methods dedicated to the digital part of the chip under test.

5.5.2 Future work

- Floor planning of the complete architecture and layout design of the complete chip so that it will cover the minimum area.
- The design of an on-chip spectrum analyzer for observing the spectral response of the ADC under different input patterns.
- Modification of the design to include the calculation of offset error, gain error and dynamic errors such as SNR, ENOB, SFDR, total harmonic distortion, etc.

References

[1] Doernberg J, Lee H S and Hodges D A 1984 Full-speed testing of A/D converters *IEEE J. Solid-State Circuits* **SC-19** 820–7

[2] Mahoney M 1987 *DSP-based Testing of Analog and Mixed-Signal Integrated Circuits* (Los Alamitos, CA: IEEE Computer Society Press)

[3] Renovell M, Azaïs F, Bernard S and Bertrand Y 2000 Hardware resource minimization for a histogram-based ADC BIST *Proc. 18th IEEE VLSI Test Symp.*

[4] Azaïs F, Bernard S, Bertrand Y and Renovell M 2000 Towards an ADC BIST scheme using the histogram test technique *Proc. IEEE European Test Workshop Cassias (Portugal)* pp 53–8

[5] Azaïs F, Bernard S, Bertrand Y and Renovell M 2001 Implementation of a linear histogram BIST for ADCs *Proc. Design, Automation and Test in Europe, Conf. and Exhibition (Munich)* pp 590–5

[6] de Vries R, Zwemstra T, Bruls E M J G and Regtien P P L 1997 Built-in self-test methodology for A/D converters *Proc. European Design and Test Conf. (ED and TC)* pp 353–8

[7] Ehsanian M, Kaminska B and Arabi K 1996 A new digital test approach for analog-to-digital converter testing *Proc. 14th VLSI Test Symp.* pp 60–5

[8] Huang J-L, Ong C-K and Cheng K-T 2000 A BIST scheme for on-chip ADC and DAC testing *Proc. Design, Automation and Test in Europe Conf. and Exhibition 2000* pp 216–20

[9] Arabi K and Kaminska B 1997 Efficient and accurate testing of analog-to-digital converters using oscillation-test method *Proc. European Design and Test Conf. (ED and TC)* pp 348–52
[10] Wang Y-S, Zhang J-L, Yu M-Y and Xiao L-Y 2007 A novel oscillation-based BIST for ADCs *7th Int. Conf. on ASIC, 2007. ASICON 2007* pp 1010–3
[11] Erdogan E S and Ozev S 2007 An ADC-BIST scheme using sequential code analysis *Proc. Design, Automation and Test in Europe Conf. and Exhibition* pp 1–6
[12] Ahmed I 2010 *Pipelined ADC Design and Enhancement Technique* (Dordrecht: Springer)
[13] Da Gloria Flores M, Negreiros M, Carro L, Susin A A, Clayton F R and Benevento C 2005 Low cost BIST for static and dynamic testing of ADCs *J. Electron. Test.* **21** 283–90
[14] Jiang S and Do M A 2008 An 8 bit 200-MS pipelined ADC with mixed-mode frontend S/H circuit *IEEE Trans. Circuits Syst. I: Regular Papers* **55** 1430–40
[15] Barua A 2014 *Analog Signal Processing: Analysis and Synthesis* (New Delhi: Wiley)
[16] Walt K 2004 *Analog to Digital Converters* (Norwood, MA: Analog Devices)
[17] Nicholas H T, Samueli H and Kim B 1988 The optimization of direct digital frequency synthesizer performance in the presence of finite word length effects *Proc. 42nd Ann. Freq. Control Symp. USERACOM (Ft. Monmouth, NJ)* pp 357–63
[18] Lu A K, Roberts G W and Johns D A 1993 A high quality analog oscillator using oversampling D/A conversion techniques *ISCAS '93 Proc. (Chicago, IL)* pp 1298–301
[19] Lu A K, Roberts G W and Johns D A 1994 A high-quality analog oscillator using oversampling D/A conversion techniques *IEEE Trans. Circuits Syst.-II: Analog Digit. Signal Process.* **41** 437–44
[20] Morris Mano M 2002 *Digital Logic and Computer Design* (Delhi: Prentice Hall)
[21] Kabisatpathy P, Barua A and Sinha S 2005 *Fault Diagnosis of Analog Integrated Circuits* (Berlin: Springer)
[22] Arabi K and Kaminska B 1997 Parametric and catastrophic fault coverage of analog circuits using oscillation-test methodology *IEEE VLSI Test Symp. (Monetrary)* pp 166–71
[23] Behzad R 2002 *Design of Analog CMOS Integrated Circuit's* (New York: McGraw-Hill)
[24] Allen P E and Holberg D R 2002 *CMOS Analog Circuit Design* (Oxford: Oxford University Press)
[25] Roberts G W and Lu A K 1998 *Analog Signal Generation for Built-In-Self-Test of Mixed-Signal Integrated Circuits* (Dordrecht: Kluwer)
[26] Stroud C E 2002 *A Designer's Guide to Built-In Self-Test* (Dordrecht: Kluwer)
[27] Sumanen L, Waltari M, Hakkarainen V and Halonen K 2002 CMOS dynamic comparators for pipeline A/D converters *Proc. of IEEE Int. Symp. on Circuits and Systems. (Phoenix-Scottsdale, AZ, 26–29 May 2002)* pp 157–60

Lightning Source UK Ltd.
Milton Keynes UK
UKHW050203270620
365617UK00003B/52

9 780750 317306